"A startling, disturbing and important piece of journalism . . . Engrossing and essential reading."
—Ron Slate, *On the Seawall: A Literary Website*

"Baker is neither obsessive nor alarmist. She calmly presents two decades' worth of critical research into the science and industries behind leading chemical culprits such as phthalates, pesticides, and PFOAs. In an appendix, she outlines the reasonable, manageable steps she's taken to detox her own home, body, and lifestyle." —*Plenty*

"This is it: The book that finally chronicles the chemical invaders tainting us and the environment . . . Any one of the chapters focusing on particular toxins (in weed killers, beauty products, cookware and computers) deserves an outraged movement." —*E, The Environmental Magazine*

"Nena Baker . . . gets her blood tested and finds out she's positive for more than three dozen toxic substances—including DDT (banned 36 years ago). This opens her investigation into our country's long history of better living through chemistry, and the price we're paying now." —*O, The Oprah Magazine*

"[Baker's] profile of UC Berkeley biologist Tyrone Hayes is particularly engaging, for he is not only a star researcher and teacher but also an impassioned advocate who was opposed by the chemical industry." —Steve Heilig, *San Francisco Chronicle Book Review*

"Important . . . Baker's is one of the growing number of voices shouting for reform and environmental cleanup."
—Donna Chavez, *SheKnows*

"Chock full of chemical history and the politics that surround it . . . [Baker] not only confronts the major issues head-on, she tells a readable story and even throws in some manageable chemistry. No easy task."
—Lisa Frack, *Enviroblog*

"Baker does a good job of tracing the Faustian bargain we have unwittingly made as a society . . . Baker is at her best as she exposes the feebleness of the most important law governing dangerous chemicals in the United States, the Toxic Substances Control Act of 1976, and the Environmental Protection Agency's anemic attempts to reform it . . . The book serves as a useful demonstration of how the PR muscle of a powerful industry can browbeat its regulators into paralysis."   —Chris Lydgate, *Lake Oswego Review*

"This important book will make it impossible to ignore the inconvenient truths about products we use every day."
—Diana Zuckerman, Ph.D., President of the National Research Center for Women and Families

"Nena Baker makes an exciting and eye-opening contribution to the growing public awareness of environmental health . . . Be astonished. Send *The Body Toxic* to everyone you care about."
—Sandra Steingraber, author of *Living Downstream: An Ecologist Looks at Cancer and the Environment*

Nena Baker

THE BODY TOXIC

Nena Baker is a former staff writer for *The Arizona Republic*, *The Oregonian*, and United Press International. Her award-winning investigation of Nike's Indonesian factories in the early 1990s led to numerous improvements for workers. She is a graduate of Lewis & Clark College and lives in Portland, Oregon.

How the Hazardous Chemistry of Everyday Things
Threatens Our Health and Well-being

Nena Baker

North Point Press
A division of Farrar, Straus and Giroux
New York

North Point Press
A division of Farrar, Straus and Giroux
18 West 18th Street, New York 10011

Copyright © 2008 by Nena Baker
All rights reserved

Printed in the United States of America
Published in 2008 by North Point Press
First paperback edition, 2009

The Library of Congress has cataloged the hardcover edition as follows:
Baker, Nena, 1959–
    The body toxic : how the hazardous chemistry of everyday things
threatens our health and well-being / Nena Baker.
        p.   cm.
    Includes bibliographical references and index.
    ISBN: 978-0-86547-707-0 (hardcover : alk. paper)
    1. Toxicology—Popular works.   2. Environmental toxicology—Popular
works.   I. Title.

RA1213.B25 2008
615.9—dc22

                                                            2008002838

Paperback ISBN: 978-0-86547-746-9

Designed by Debbie Glasserman

www.fsgbooks.com

P1

For Patti, with love and appreciation

Contents

Introduction: Coming Clean   3

1   A Chemical Stew: Body Burden   11

2   Chemicals We've Loved:
    Consumer Conveniences   33

3   Kermit's Blues: Atrazine and Frogs   57

4   What Price Beauty? Phthalates and You   88

5   Up in Flames: Polybrominated
    Diphenyl Ethers   112

6   The Goods on Bad Plastic: Bisphenol A   141

7   Out of the Frying Pan and onto the Paper:
    Perfluorinated Chemicals   163

8   Reaching Ahead: New Policies   190

Epilogue: My List and Beyond   212

   Appendix 1: It's All About You   217

   Appendix 2: Environmental and Public-Health
             Groups That Get It   225

   Appendix 3: Learn More from Government
             Sources   229

   Notes   233

   Acknowledgments   257

   Index   259

The BODY TOXIC

INTRODUCTION: COMING CLEAN

One evening in early 2003 I was perusing the pages of *The New York Times* when an article about a study from the Centers for Disease Control and Prevention caught my eye. The CDC said that with the help of technology that detects an increasing number of chemicals in human blood and tissue, it would be releasing new measurements, called biomonitoring, every few years. This bit of news, which netted only a few paragraphs in the paper, seemed to me a much bigger event than that. The discomfiting fact that small amounts of chemical pollutants—including substances used in everyday products—pulse through my body and everyone else's expands the very definition of pollution. No longer can we think of pollution as an external insult that affects only the environment. Our bodies bear the burden, too.

The CDC's promise of regular biomonitoring reporting

raises intriguing questions. Should we be worried about
the effects of these pollutants on our health? Can everyday
items be responsible for the chemicals inside us? Don't
regulators already make sure we're safe from daily doses of
hazardous substances? I started digging and soon discovered a situation unlike any I had encountered in all my
years as an investigative reporter. It inspired me to leave
daily journalism after two decades to write *The Body
Toxic*.

In short, the United States does not have a viable means
to keep its 300 million citizens safe from untold chemical
hazards in the things we use and buy day in and day out.
As a result of this failure, chemicals that can interfere with
the body's reproductive, developmental, and behavioral
systems are freely used in everything from plastics, soaps,
and toys to food, food wrappers, clothing, and carpeting. Hundreds of peer-reviewed studies show that these
endocrine-disrupting chemicals, which throw off the
body's hormone system in various ways, cause lab animals
to exhibit disorders and diseases that are on the rise in humans. The ghoulish list includes cancers of the breast, tes-
ticles, and brain; lowered sperm count; early puberty;
endometriosis and other defects of the female reproductive system; diabetes; obesity; attention deficit disorder;
asthma; and autism. Getting back to the CDC's biomonitoring work: the chemicals not only contaminate our
homes, offices, and vehicles but are also inside us at levels
that, in a few cases, are equal to or uncomfortably close to
the amounts that cause harmful effects in lab animals.

Everyone is affected by the phenomenon. But unlike
global warming, this public-health crisis has not been
blasted into the blogosphere or made into a movie by Al

Gore. At least, not yet. But awareness of toxic chemicals in everyday products began to take hold during the four years that I worked on this book. Important new research studies grabbed headlines. New chemical regulations in the European Union and Canada emphasized consumer safety over corporate profits. Product scares brought home the danger of everyday exposures to toxics.

As I was finishing these pages, parents across the land crept through the night stealing toys from their babies because of lead-safety issues involving millions of popular playthings. Meanwhile, worried pet owners besieged the Food and Drug Administration with eighteen thousand phone calls after an outbreak of animal deaths from melamine-laced food. The problem products shared a common place of origin: China. As the recalls mounted, so, too, did cries that Chinese manufacturers do not live up to U.S. lead-safety standards introduced in the 1970s, when the United States banned lead in paint because of its potent toxicity to the brain and central nervous system. Prohibitions on leaded solder in plumbing and food cans soon followed. And in the early 1990s, the United States completed its phaseout of leaded gasoline. The resulting reductions in the blood lead levels of U.S. adults and children, as tracked by the CDC, are considered one of the great success stories of public health.

The toy recalls and pet-food fiasco soured many consumers on Chinese-made products, with nearly half of all Americans, according to a Harris poll, saying they would avoid any type of item fabricated there. (Quite a task given that the U.S. Census Bureau says we import more goods from China than from any other country.) Meanwhile, the U.S. Senate called for mandatory safety testing of all toys sold in the United States because voluntary self-policing by manufacturers, importers, and retailers and spot checks by a woefully understaffed and underfunded Consumer Product Safety Commission simply don't work. In April 2008, Washington State recognized this fact by enacting the toughest toy safety standards in the nation: toys containing unacceptable levels of lead, cadmium, and plasticizers called phthalates will be outlawed as of July 2009. Consumers should be concerned about the dearth of safety standards in developing countries, where the vast majority of U.S. products are made. And Congress has responded with sweeping reforms to ensure that children can play with any toy without sending their parents scurrying for lead-test kits or chemistry textbooks. The trouble with toxics, however, goes beyond where a product comes from and a few substances of long-standing worry.

Inventory your house and you'll see why. Televisions are treated with flame retardants; furniture and carpets are coated with stain fighters; food containers take form from plasticizers; plastic toys—even those without lead paint—may be molded from polyvinyl chloride (PVC); and the bathroom shelf brims with chemical-laden personal care items. These products and treatments are problematic for a variety of reasons, including their potential to contribute to human exposures to hormone-mimicking substances. As with lead, children are the most susceptible to potential lifelong impacts from these toxics because their metabolism and behaviors expose them disproportionately.

The chemical industry insists everyday exposures to endocrine-disrupting substances are inconsequential to humans, young or old, because the amounts are too minuscule to matter. Such assurances are backed up with studies that, with rare exceptions, are funded by the industry itself. I don't want to give away too much here about what you'll discover in the book, but suffice it to say, the way chemical regulations work benefits the industry at the expense of the public. Yet manufacturers did not cook up our chemical stew all by themselves. To suggest so overlooks gross congressional failures: for more than three decades, our elected leaders, both Democrats and Republicans, have let stand the notoriously weak and ineffectual Toxic Substances and Control Act of 1976, which governs the use of some 82,000 chemicals.

Through these pages, I will show just how spectacularly this landmark legislation falls short of what it was intended to do: protect public health and the environment. Indeed, the federal toxics law *discourages* chemical companies from knowing and sharing hazard and exposure information—the two variables that must be known in order for regulators to conduct risk assessments, according to Dr. Lynn Goldman, a pediatrician and professor of environmental health sciences at the Johns Hopkins Bloomberg School of Public Health. From 1993 to 1998, Goldman served as assistant administrator of the Environmental Protection Agency's Office of Prevention, Pesticides and Toxic Substances, where she set up voluntary programs to generate data from chemical manufacturers. Goldman is the first to concede that the honor system hasn't worked.

"As soon as [chemical manufacturers] identify a new problem with a chemical, then that chemical becomes vulnerable to regulation," says Goldman. "And so if you were sitting there worried about protecting shareholder value, would the first thing on your mind be to go out and find more problems with your product that will then subject it to more regulation? It would not, because the more regulations, the less likely your customers are to want to purchase the chemical from you. And so in the way the laws are structured, there's a perverse incentive not to look. The financial incentive is that as long as you don't look, if you have no data about hazards, no data about exposures, then there is no risk assessment and then there is no risk, which is, of course, not actually true. But that is, in effect, how it works."

Through the years, authorities no less than the National Academy of Sciences, the U.S. General Accounting Office, the congressional Office of Technology Assessment, and the U.S. Government Accountability Office (GAO) have weighed in on the inadequacies of federal regulations governing the use of toxic substances on inventory with the EPA. In its most recent report, the GAO recommended that Congress strengthen the toxics act to give the EPA the authority it lacks to do what most citizens assume it already does—assess and control chemicals that cause harm. Where endocrine-disrupting chemicals are concerned, Congress instructed the EPA more than a decade ago to begin a screening program in order to identify substances that may interfere with biological processes and change the way the body functions. That was in 1996, just as the theory of endocrine disruption was emerging. To date, the EPA has spent some $70 million but has yet to identify even *one* substance for chemical manufacturers to begin screening. The House Committee on Oversight and Government Reform is now riding herd on the EPA to take "adequate and timely steps to protect the American public from dangerous endocrine-disrupting chemicals." Wasted time and long delays are the rule when it comes to toxics testing, putting all Americans in harm's way.

From a consumer's point of view, the situation is equally appalling at the FDA, which oversees $1 trillion a year of food, drugs, medical devices, and cosmetics. A scathing 2007 report prepared by members of the agency's own science advisory panel concluded the FDA is suffering from serious deficiencies that put American lives at risk. Noting that the FDA's resources have tightened as its workload and need for scientific sophistication have soared, the report said, "This imbalance is imposing a significant risk to the integrity of the food, drug, cosmetic, and device regulatory system, and hence the safety of the public." When a crisis erupts involving cosmetics, food, or drugs, the FDA cannot adequately respond. One example: in April 2008, the FDA agreed to review bisphenol A, the backbone of polycarbonate plastic, only after Canadian regulators moved to ban it in baby bottles and a U.S. House committee began investigating.

Acknowledging the breathtaking scope of the challenges faced by the FDA, then commissioner Andrew C. von Eschenbach offered this alarming assessment of the agency: "The simple truth as I see it today is that the FDA of the twentieth century is not adequate to regulate the food and drugs of the twenty-first century . . . FDA was created one hundred years ago because change had created peril along with promise, and today FDA must be re-created because the peril and promise from these products is now even greater." As the FDA founders, scant resources hamper the Consumer Product Safety Commission. Full-time positions dwindled from 978 in 1980 to a paltry 400 staffers in early 2008. Under the George W.

Bush administration, the CPSC took a hands-off approach that veered further than ever from the agency's mandate to regulate products that present an unreasonable risk. As an example of the agency's nonchalance, Acting Chairman Nancy Nord stated that Congress never intended for the agency to inspect all consumer products. It would, she said, be "unrealistic, not to mention the drag of such an effort on global commerce, our economy and ultimately higher product costs."

Buyer beware, indeed. Congress moved to establish clear safety standards for toys and other children's items in the summer of 2008 by passing the Consumer Product Safety Improvement Act. But the fact remains: for more than three decades, the chemical industry, with the complicity of our elected leaders, has kept us in the dark about the toxicity of everyday substances and successfully resisted policy efforts that would better protect the public. It's high time for chemical makers and Congress to come clean.

A CHEMICAL STEW: BODY BURDEN

The turnoff to the tiny hamlet of Bolinas is unmarked from California Highway 1 as it twists along Pacific Ocean headlands one hour north of San Francisco. Every time highway crews put up a sign pointing to Bolinas, the locals take it down. A building moratorium enacted in 1971 preserves Bolinas much as it was during its counterculture heyday: a colony of 1,560 artists, writers, healers, and activists intent on safeguarding their bohemian community from commercial encroachment. While minimansions and new subdivisions dot nearby Stinson Beach, Bolinas still looks like it did when Richard Nixon was in the White House and Bill Clinton inhaled.

Downtown boasts a grocery store with more free-range dogs loitering outside than patrons shopping inside, a restaurant that serves the freshest ingredients from nearby farms, and a gas station with a bed-and-breakfast above it.

Victorian houses and weathered clapboard cottages rim the shore of Bolinas Lagoon, a haven for pelicans and a regular pit stop for migratory birds navigating the Pacific Flyway. Living costs have gone up in Bolinas, but local sensibilities and the pristine landscape have stayed the same.

Twenty years ago, Sharyle Patton discovered the town and fell in love with it. "I used to come out to Bolinas and play music," said Patton, a pianist and singer who studied at the Berklee College of Music in Boston. That led to playing bluegrass and jazz for a living. She met her husband, Michael Lerner, in Bolinas. He founded and directs Commonweal, an alternative-medicine think tank and cancer healing center that occupies a former RCA transmission site overlooking the Pacific in Bolinas, where the couple lives.

Approaching the age of sixty, Patton has the trim build and spirited glow of a woman who pays attention to diet and exercise. "It's easy to eat organic in Bolinas," said Patton, who also takes advantage of miles of beaches right outside her door and nearby hiking trails that crisscross breathtaking vistas in Point Reyes National Seashore. She was raised on a Colorado ranch and she likes to be outdoors. The bungalow she shares with her husband came with a spectacular garden. Patton enjoys tending the previous owner's legacy, adding more color and texture to the garden every year.

She's always taken good care of herself, avoiding the pitfalls of drugs, booze, and tobacco that plagued others of her generation, especially fellow musicians. And it shows. She stands straight, which makes her look even taller than her five feet eight. A short tousle of blond hair frames blue eyes that twinkle and a wide, slightly lopsided smile. Patton displays the energy of a woman half her age as an activist on issues of health and the environment. In 2001 in Stockholm, as a leader of a network of 350 non-governmental organizations from around the world, Patton helped guide the UN's Persistent Organic Pollutants treaty, which calls for the worldwide elimination of a "dirty dozen" list of chemical contaminants considered among the world's most hazardous.

Intellectually, she understands as well as anyone the ubiquitous nature of chemical pollutants. But she didn't expect the emotional jolt she felt when she learned that her body was polluted with traces of 105 chemicals linked in animal studies to a list of devastating health effects including cancer, disruption of the hormone system, birth deformities, and neurological impairments. "I don't live next door to a refinery or an incinerator or some kind of factory," said Patton, whose blood and urine were screened for chemical pollutants after she volunteered for a study conducted by Mount Sinai School of Medicine in New York. "I've been careful and it hasn't made a bit of difference in terms of the chemicals that are in my body."

It turns out that what's in Patton is in every one of us, too. Unlike our forebears, everyone everywhere now carries a dizzying array of chemical contaminants, the by-products of modern industry and innovation. These toxic substances accumulate in our fat, bones, blood, and organs, or pass through us in breast milk, urine, feces, sweat, semen, hair, and nails. Scientists studying pollutants in people—including researchers at the Centers for Disease Control and Prevention (CDC) in Atlanta—call the phenomenon "chemical body burden." It is the consequence of womb-to-tomb exposures to substances so common in our daily lives that we never stop to consider them.

That water-repellent jacket you're wearing? It got that way because of a chemical called perfluorooctanoic acid (PFOA), which is used to make the fluoropolymer membranes needed to impart the extra utility. As of this writing, the Environmental Protection Agency, which has asked manufacturers to voluntarily reduce emissions of PFOA, is debating whether to officially describe the substance as "likely to cause carcinogenicity" in humans.

That cute yellow bath toy your child or grandchild loves to chew? It's likely to contain plasticizers known as phthalates (pronounced "THAL-ates"), which are part of a large family of industrial chemicals linked to impaired sperm quality in animals.

That TV you spend hours in front of? It's probably made with a neurotoxic chemical flame retardant known as polybrominated diphenyl ether (PBDE), which is showing up in the breast milk of U.S. mothers at rates one hundred times the average found in European studies. In 2003, California followed the lead of the European Union and became the first state to ban two types of PBDEs. Other states have followed. But the most common type of PBDE—and the one found in televisions—is still in wide use. Scientists are worried that PBDEs disrupt the developing thyroid system and could cause developmental deficits.

"It's overwhelming what we're exposed to," said Jane Houlihan, vice president of research for the Environmental Working Group (EWG), a Washington, D.C.–based public-advocacy organization that partnered in several biomonitoring studies in an effort to raise awareness of chemical body burden and the need for more research about the health effects of low-level exposures. "Every day we get a fresh flush of chemicals."

No place—and no one—is immune. The most persistent chemical contaminants are carried across oceans and continents by water and air. Like grasshoppers, they lift into the atmosphere, then glide back to earth, moving from warmer climates to colder climates and settling thousands of miles from a contamination source. They're fat soluble and they bio-magnify, increasing in concentration as they move up the food chain. They cross the placenta, so babies receive their first exposures in the womb.

"Not long ago scientists thought that the placenta shielded cord blood—and the developing baby—from most chemicals and pollutants in the environment," wrote the authors of a 2005 study sponsored by EWG that measured an average of two hundred chemicals in the umbilical cord blood of ten newborn American infants. "But now we know that at this critical time when organs, vessels, membranes and systems are knit together from single cells to finished form in a span of weeks, the umbilical cord carries not only the building blocks of life, but also a steady stream of industrial chemicals, pollutants and pesticides that cross the placenta as readily as residues from cigarettes and alcohol."

In the United States, our chemical neighborhood includes more than eighty thousand industrial substances registered for commercial purposes with the EPA. About ten thousand of these chemicals are widely used in everything from clothing, carpeting, household cleaners, and computers to furniture, food, food containers, paint, cookware, and cosmetics. But *the vast majority of them have not been tested for potential toxic effects* because the U.S. Toxic Substances Control Act (TSCA) of 1976 does not require it. And the news gets shockingly worse: the EPA cannot take any regulatory action regarding a suspected harmful substance until it has evidence that it poses an "unreasonable" risk of injury to human health or the environment. The barriers to action are so high that, according to a 2005 report by the Government Accountability Office, the EPA has given up trying to regulate chemicals and instead relies on the chemical industry to act voluntarily when concerns arise. These stunning policy failures have not been rectified in more than three decades.

Indisputably, chemicals have helped raise our living standard and make our lives easier and safer. Think of the conveniences of plastic food-storage containers, stain-resistant carpeting, sleek personal computers, and fast-cooking microwaves. Think of the security of fire-resistant materials, clean-water supplies, stronger-than-steel bulletproof vests and nylon seat belts. Who can argue with the American Chemistry Council (ACC), a trade organization that represents Dow, DuPont, and hundreds of other chemical companies, when it suggests, in a new advertising campaign, that chemicals are "essential2life"? Putting the unctuous text-message-style grammar aside, the slogan speaks volumes about the importance of the $664-billion-a-year U.S. industry and the seventy thousand products for which it supplies raw materials. It's not that chemicals are bad per se, and it would be preposterous for even the most ardent environmentalist to suggest such a notion. It's that costly societal problems often arise because we know so little about so many chemicals. And in the time it takes to learn what harm a substance is doing—to people, to animals, to places—the genie is long out of the bottle.

Examples originate from all over the periodic table: heavy metals such as lead, mercury, and cadmium that steal precious brain function; potent organochlorine substances such as the pesticide DDT, the industrial insulators known as polychlorinated biphenyls or PCBs, and the man-made chemical by-products called dioxins that have been contaminating the environment for decades; and halogenated compounds such as ozone-depleting chlorofluorocarbons. Even when regulators intervene to ban or restrict their use, some of these substances show up in people years after they've been proved unsafe.

My own body burden analysis, performed by a laboratory in Manchester, England, at a cost of nearly $2,000, confirmed I am a poster child for the era of better living through chemistry. On a beautiful fall day, I visited my doctor's office in Portland, Oregon, where a lab technician drew six tubes of my blood. She centrifuged the containers in order to separate the wine-colored cells and platelets from the wheat-colored serum. Then I carefully boxed the tubes in dry ice to preserve them for transport and drove to a FedEx office. Eight weeks later, an e-mail from Scientific Analysis Laboratories settled in my in-box. The attached spreadsheet showed that my body contained traces of at least three dozen persistent toxic chemicals, including DDT and PCBs. Both of these toxics were banned in the United States and Europe thirty years ago when I was a teenager, but their persistence in the environment means they are still part of the food web, contributing to my body burden—and everyone else's. These two chemicals and their metabolites, which are the substances formed as the original compound breaks down, are routinely detected in the population by the CDC's biomonitoring program and others around the world.

The good news is that human exposure levels are decreasing, confirming the wisdom of banning these long-lived pollutants. But my blood also contained traces of PBDES, which are building up in North Americans at rates so astonishing that some researchers refer to these chemicals as "the PCBs of the twenty-first century." These flame retardants have been widely used in North American consumer products ranging from computers to office chairs. The U.S. manufacturers of PBDEs agreed to stop producing two forms of the chemical in 2004. But production continues on the most widely used type, and experts predict that the PBDE-containing products that fill homes and offices will contaminate people for many years to come. As a result, scientists worry that PBDEs have the potential to create a new public-health crisis. "It's scary to see chemicals that behave like PCBs rising in the environment and in our bodies," said Dr. Gina Solomon, a senior scientist for the Natural Resources Defense Council and assistant professor of medicine at the University of California at San Francisco.

Similarly, perfluorinated chemicals (PFCs), an ingredient in the fluoropolymers essential for making nonstick coatings, water-repelling fabrics, oil-and grease-resistant food packaging, and firefighting foam, are ubiquitous in people and the environment. Most levels of the three dozen chemicals and metabolites measured in my body were about average compared to results from other body burden studies. The exception, however, was perfluorooctane sulfonate or PFOS, a chemical once used in a broad range of industrial, commercial, and consumer products, including 3M's Scotchgard line for carpet, furniture, and clothing. PFOS has been shown to cause cancer and developmental problems in lab animals. At 77.2 parts per billion, my exposure to PFOS was higher than the maximum levels detected for nonoccupationally exposed adults in three comparison studies, and more than four times greater than the average 18.4 parts per billion for adult women reported in 2007 by CDC researchers. In 2000, at a cost of $200 million, 3M announced it would quit making PFOS and that it was reformulating the Scotchgard line using a closely related chemical alternative. The EPA proposed a "significant new use rule" in October 2000 to limit new uses of PFOS. Around the same time, the agency expanded its investigation to include PFOA, a related chemical used in making Teflon cookware coatings and water-resistant fabrics. In early 2006, the EPA announced a deal intended to reduce PFOA releases and product-content levels with the eight companies, including DuPont, that manufacture the substance in the United States. These companies committed to work toward elimination of all sources of PFOA exposure no later than 2015. But there's much more to the story of perfluorinated chemicals and their potential for harm.

I have no idea why I have higher exposure to PFOS than most other adults. Biomonitoring studies can't tell scientists anything about the source of exposures or when they occurred. So I'm left to wonder if my exposures came from products or from the food chain. In either case, it was largely unavoidable. I came in contact with this extremely persistent, bioaccumulative chemical through the course of my daily living. And because of the substance's remarkable ability to survive—it does not biodegrade or break down in the environment and it is slowly metabolized in animals and humans—it will be in my body for many years, even in the unlikely event that I never have another exposure. And I have no idea yet what this might mean for me healthwise.

Scientists sometimes refer to perfluorinated chemicals, in use since the 1950s, and brominated flame retardants, which were introduced in the 1970s, as emerging chemicals of concern. When I think about that term, it makes me wonder if we've learned anything from our earlier experiences with other persistent substances. The story lines of the "emerging" contaminants are strikingly similar to DDT and PCBs: persistent substances used widely for years before their toxicity was truly understood. DDT and PCBs are now known as "legacy" chemicals because of their continued presence in the environment and in people. Because of gross inadequacies in the U.S. toxics law, perfluorinated chemicals and brominated flame retardants are destined to become the legacy substances of tomorrow.

Certainly, these substances are as stubborn to scrub from the environment as PCBs, which were banned by the EPA in 1978. We continue to ingest PCBs when we eat fish, meat, and dairy products, which is why PCBs were included in the 2001 UN chemicals treaty for worldwide elimination that Sharyle Patton helped to craft. Research shows that contaminated food is the pathway for many air- and waterborne toxics, including pesticides; pollutants called dioxins and furans generated from waste incineration and fuel combustion; methylmercury from coal-fired power plants; and cadmium from mining and industrial operations.

But other contaminants found in Patton most likely come from direct contact with a source through breathing or touching. For example, phthalates are ingredients in cosmetics, scented soaps, deodorants, and plastic food containers. CDC data show that exposure to phthalates is widespread throughout the population, and women of childbearing age have the highest levels. For Patton, who won her study's sweepstakes for most types of PCBs in her body, learning about her body burden profile raised some troubling health-related questions. Which of the 105 chemicals measured in her blood and urine should she be most worried about? What role had her chemical load played in her inability to have children? How will her body burden affect her in old age? As it stands, physicians and researchers need more data to suss out the substances of greatest concern. And distinguishing a cause-and-effect relationship between environmental exposures and specific diseases or dysfunction is one of the most daunting challenges of epidemiology. "We're just beginning a decade or two of science that will help answer those questions," said Dr. Michael McCally, a professor of community and preventive medicine at Mount Sinai and a coauthor of the study in which Patton participated. It is also why individual body burden screening, which at the time of this writing costs up to $5,000 and is conducted at only a handful of laboratories, is a long way from being used as a diagnostic tool in the same vein as, say, allergy tests or Pap smears. Nevertheless, scientists see great promise in measuring human body burdens.

"Biomonitoring is not just this little field over here that

gives everybody new and special information," said Dr. Jim Pirkle, who runs the world's most advanced biomonitoring laboratory in a new three-story building on the sprawling campus of the CDC just northeast of Atlanta. "It's exciting because it contributes in a major way to every single step that helps us prevent environmental disease." In order for epidemiologists to reliably link environmental pollutants with health problems, they need to know who has been exposed and at what level. In the past, this information came from questioning people about their exposures, which often produced sketchy results. Alternatively, researchers might measure exposure levels found in an individual's surroundings, but that did not tell them a thing about what actually was absorbed by the body. Biomonitoring neatly resolves these problems by measuring the actual levels of environmental chemicals in human tissues and fluids. Pirkle, a physician who has been with the CDC since the late 1970s, describes it as the "gold standard" for assessing exposure to chemicals. "Measuring what gets into people is just a quantum step in terms of the value of the information," he said. And with that information, researchers can begin to accurately assess health risks, develop and apply interventions, and finally, figure out if they are on the right path to reducing human exposures.

The mass spectrometers needed to measure traces of chemicals and pesticides in human samples of blood and urine are big beige machines that, to the untrained eye, look like something you might find at your neighborhood Kinko's. But chemists drool over the sophistication of the setup and the cutting-edge work coming from the CDC's lab. Commanded by researchers and technicians in clean suits seated at computer keyboards, these exquisitely sensitive machines spin out findings that show how the substances we rely on to make life easier and more convenient have the unintended effect of polluting us.

The CDC is tracking dozens of pesticides and chemical compounds we encounter in daily life in the largest ongoing study of chemical exposures ever conducted on humans. In its most recent survey, its third since it began sampling U.S. citizens for chemical pollutants in 2001, the CDC measured 148 substances. With each survey, the CDC is adding more chemicals to the list. The next CDC report is due in late 2008. Advocacy organizations conducting their own biomonitoring studies have measured some substances yet to appear in the CDC reports. However, for every chemical and pesticide measured, there are hundreds more for which the means have not yet been developed to screen for their presence in humans.

While biomonitoring studies provide a much more accurate picture of our chemical body burden, limitations remain. The studies don't tell researchers the source of an exposure, how long a substance has been in the body or, most important, what effects, if any, a substance is having on human health. "That's the $64,000 question," said Dr. Linda Birnbaum, director of experimental toxicology for the EPA. "We really need more research to understand whether the levels we're finding could be associated with adverse health effects." Hundreds of scientific studies, based on data collected by the CDC, are under way at government, academic, and industry research centers to help understand health outcomes. Meanwhile, after two years of contentious debate in the California legislature, Governor Arnold Schwarzenegger signed a bill in 2006 that establishes a voluntary statewide biomonitoring program. The goal is to produce statistically significant findings on environmental chemical exposure among Californians that will help guide the state's public-health priorities.

So how might that work? An example the CDC's Pirkle shared with me from the early days of biomonitoring explains how valuable it can be in setting public policy. In 1976, when the United States began taking lead out of gasoline, the EPA predicted, based on mathematical models, that levels in people would drop from an average of 1,580 parts per billion to 1,520 parts per billion. However, a CDC biomonitoring study showed that lead level reductions were actually ten times greater than what the EPA had predicted.

"This was a huge, huge deal because in 1981 EPA made a proposal to increase the amount of lead in leaded gasoline," said Pirkle, whose blue eyes twinkle behind his aviator-style glasses as he relishes the story. "The EPA made this proposal because of what their mathematical models told them. They knew that lead was an inexpensive octane booster, and they figured they could increase lead and it wouldn't be a problem for people. I'm not blaming them. That's what their models told them."

However, when the EPA put out a call for comments on the proposal, Pirkle and his research team raced north to Washington to present the EPA with the biomonitoring data showing the huge decrease in lead levels. "We said, 'Look at this. When you take the lead out, it actually makes a much bigger difference than you think.' " Instead of letting the lead back in, the EPA moved in the other direction to more quickly eliminate lead from gasoline. "They changed their minds and completely turned the whole thing around because we had the data," he said. "It was really a dramatic thing."

As the ability to pinpoint pollutants in people advances, it creates tension between those who see no value in measuring traces of toxics that are below the levels of any known health risks and those who see the pervasiveness of the exposures as a fundamental right-to-know issue, if not a powerful clarion for more precautions. Dr. Richard Jackson, the former director of the CDC's National Center for Environmental Health, which oversees the biomonitoring lab, told me he came under tremendous pressure in 2001 when the CDC's first national exposure survey was released. "I took a fair amount of criticism for disseminating the report without putting it through some kind of extensive risk assessment," he said. "But I resisted that very strongly, not because I was antiscientific, but because I wanted the larger community and the research community to have it in their hands and use this data the way a doctor would use lab data in making decisions about a patient. The complaint from chemical manufacturers was that the report was just going to scare people. No, you never scare people with real information. You scare them with no information or bad information."

For years now, it's been well understood that the risk of developing an environmentally related disease or disability depends on a combination of environmental exposures, genes, age, sex, nutrition, and lifestyle. And we all know that lifestyle choices such as diet, smoking, alcohol use, exercise levels, and UV exposures can influence our genetic susceptibility to health problems and diseases such as high blood pressure, heart disease, or cancer. But some of the newest research is suggesting something about low-level exposures to certain substances that is not so well understood. Since the beginning of the chemical revolution, manufacturers have held fast to an axiom in toxicology that "the dose makes the poison." One of the fifteenth-century fathers of modern medicine, Paracelsus, came up with the concept, which means the higher the dose, the greater the effect. In other words, trace levels of contamination like those found in body burden studies aren't worth worrying about. So, in large part, chemical manufacturers have shrugged them off. "Just because we can detect a chemical doesn't mean there's a problem," said Carol Henry, vice president for industry performance programs for the ACC. As a result, regulatory toxicology, often conducted by chemical companies themselves, focuses on adult health effects such as cancer and obvious signs of organ damage at high-dose exposures. In contrast, research toxicology, primarily conducted at independent university and government research centers, studies low-dose exposures, which cause more subtle alterations of bodily systems that regulate such things as organ function, sexual development, behavioral cues, intelligence, and reproductive systems.

Mounting evidence suggests that tiny amounts of certain chemicals, both natural and synthetic, can wreak havoc with hormones that shape our bodies and behavior over a lifetime. These chemicals are called endocrine disruptors because of the tricks they play with the complex physiology that controls basic systems of the body from fetal development through adulthood. Research shows that endocrine disruptors may pose the greatest risk during prenatal and early postnatal development, even though the negative effects of that exposure may not be experienced until much later in life. All of the chemicals I explore in detail later in this book are known or suspected endocrine-disrupting substances.

A few years ago, an independent panel of experts convened by the National Institute of Environmental Health Sciences (NIEHS) and the National Toxicology Program (NTP) found there was "credible evidence" that some hormonelike chemicals affect test animals' bodily functions at measurements so low they were beneath the "no-effect" levels determined by standard toxicology tests. The state of the research was summarized in a startling 2006 fact sheet from NIEHS: "There is a large body of research in experimental animals and wildlife suggesting endocrine disruptors may cause: reduced male fertility and declines in the numbers of males born; abnormalities in male reproductive organs; female reproductive diseases including early puberty and early reproductive [declines] and increases in cancers of the breast, ovaries and prostate." According to the NIEHS, data show that exposure to some endocrine-disrupting chemicals such as bisphenol A, a controversial chemical used in the manufacture of polycarbonate plastic products including baby bottles and Nalgene bottles, may have effects on obesity and diabetes. The implications of bisphenol A use are explored in detail in Chapter 6. But among the most disturbing is evidence that bisphenol A and other endocrine disruptors may cause effects in an exposed individual's children and grandchildren.

As always, there is a crucial need for more research. No one is exposed to just one chemical at a time, which further complicates the research picture because chemical mixtures can alter the toxic properties of individual chemicals. We each have a unique—and constantly fluctuating—mixture of chemicals inside of us depending on new exposures and our ability to eliminate old ones. Some of us, because of our susceptibility and genetic makeup, may be hurt by a chemical or chemical combination that doesn't bother most other people. "We don't know how important chemical mixtures are as opposed to genetics, but we do know that chemicals can put someone over the edge in terms of developing a disease," said Houlihan of the Environmental Working Group. "Genetics make them vulnerable, but environmental exposures are the final trigger."

Researchers have already found links between chemical exposures and some portion of childhood asthma, certain cancers, and neurodevelopmental disorders. Chronic conditions of multifactorial origin have been termed "the new pediatric morbidity," and many are on the rise. Indeed, two of the world's leading researchers on brain development recently suggested in a review published in the influential medical journal *Lancet* that fetal and early childhood exposures to hundreds of chemical contaminants are causing "a silent pandemic" of brain disorders. Dr. Philippe Grandjean, adjunct professor of environmental health at the Harvard School of Public Health, and Dr. Philip Landrigan, professor of community and preventive medicine at Mount Sinai School of Medicine in New York, said the pandemic is preventable if regulators set stricter exposure standards that "recognize the unique sensitivity of pregnant women and young children."

Should the need for further research prevent us from taking responsible actions now? The European Union has said no and has taken the lead with groundbreaking legislation that requires chemical companies to prove that thousands of everyday chemicals are safe. But, as I explore much more fully later, U.S. regulators—often criticized as ineffectual public watchdogs against the chemical industry's pressure tactics—have made no such move. Nevertheless, the pressure is growing, especially as scientists continue to quantify the low-level doses of chemicals polluting people.

"The bad news is that this new rush of science, the torrent of research that is coming out every week, is telling us that we have inadvertently allowed into the environment, into consumer products, into some of the food that we eat and the water that we drink, a series of contaminants that interfere with how genes are expressed and thereby affect the biological development of organisms," explains John Peterson "Pete" Myers, a coauthor of *Our Stolen Future* and the founder and CEO of a news organization called Environmental Health Sciences. "The good news is that this wave of new science is opening up many opportunities for preventing diseases that heretofore we had no clue were linked to environmental exposures."

Ridding our bodies of these chemical invaders simply isn't possible because of their persistence and the frequency of new exposures. But there are better ways to move forward. Protecting future generations from contamination requires identifying and eliminating the most dangerous persistent, bioaccumulative chemicals and developing alternative production methods that use non-toxic materials.

On a personal level, if we learn about the sources and pathways of low-level exposures, we can try to avoid new ones. Even small changes in lifestyle—choosing to eat low-fat, organic foods, for example, or microwaving only in glass or ceramic containers—can help reduce the chances of adding to a chemical load.

Back in Bolinas, sitting in her living room on a comfortable green sofa, Sharyle Patton is very sad and angry. "Our air, water, soil, and our own bodies are being used as a chemical sewer system," she said. She is incredulous that it's easier to prevent commercial encroachment in Bolinas than it is to stop unwanted toxic substances from trespassing in her body. But it is. Even more distressing, she said, is how little is known about the health effects toxics have on our bodies and a regulatory system that enables that to happen. "At a minimum, we have a right to know what's in us," she said. Body burden studies help. Yet she worries, as we all should, about the damage done. One of the cruelest realities of body burden is that mothers and babies—the very key to our future—are most vulnerable to chemical contaminants. Through the miracles of modern chemistry and society's cravings for convenience, we've fundamentally altered, in less than a century, a biological balance that evolved over millions of years. And we don't know the full costs.

On many occasions during the research for this book, I've thought back to a rainy winter night several years ago. As I sat in a hospital cafeteria picking at my dinner, a young resident who had visited my very ill mother earlier in the day joined me at my table. At the age of seventy, my mother had pancreatic cancer and the worst prognosis possible. She was close to death. The kind young doctor and I had met only that day. As we talked, I mentioned that my mother had been treated previously for bladder cancer and breast cancer, both of which appeared to be in remission. As far as I knew, her cancers were not related to one another. "How could she be that unlucky?" I wondered. "Maybe it's not just bad luck," the doctor said, going on to ask where she lived and what kind of work she did. The doctor was thinking about her possible exposures to large amounts of toxics that might have overwhelmed her body's defense system. My mother, a lifelong city dweller who had never lived or worked around manufacturing plants, landfills, or farmers' fields, was not occupationally or accidentally exposed to dangerous chemicals. She was a homemaker—chief cleaner, cook, and chauffeur for our family of four—who lived an ordinary life for a woman of her generation.

Today, however, we are learning that ordinary exposures to everyday chemicals, especially during critical windows of fetal development, may have extraordinary impacts on our bodies and brains. In effect, certain contaminants can hijack the control of gene expression, contributing to a number of illnesses that are on the rise. The concept crystallized for me as I listened to a lecture by Pete Myers, who used a particularly compelling metaphor to explain why we should all be concerned. "Think about September 11," he said. "What happened? A small number of guys went underneath the radar screen of the defense system and through airport security without attracting attention. They went on planes, grabbed control of the steering mechanisms of the jets, and took them in directions the jets never would have gone under normal circumstances. Some contaminants are doing the same thing in our bodies. They are coming in at low levels, grabbing control of the mechanisms, and taking the organism along a developmental path it never would have taken without the presence of those factors interfering in gene expression."

I will never know if my mother's genes were, in effect, hijacked by low-level exposures to everyday chemicals that proliferated in her lifetime. What I can be certain of, however, is she gleefully embraced the advances of the twentieth century and, like many of her generation, couldn't wait for the next modern breakthrough to make life more comfortable, convenient, and safe for the people she loved. She believed in better living through chemistry. Above all, she trusted it to be true.

CHEMICALS WE'VE LOVED: CONSUMER CONVENIENCES

By the mid-twentieth century, it was abundantly clear that science ruled our worst fears and our best hopes. While anxieties festered over the potential for a nuclear war between the world's superpowers, a steady stream of new products—all made possible by the miracles of chemistry—promised peace of mind, greater well-being, and lots more fun. In July 1955, as Disneyland opened in sun-drenched Anaheim, California, on the site of a former citrus grove, a group of Nobel Prize winners ominously warned that science was giving humankind the means to destroy itself. Nuclear fallout would disastrously alter our fate. Nevertheless, the buildup of cataclysmic weaponry continued, and so did the buildup of stuff. *Time* magazine described a nation "in the heady grip of a prosperity psychology." Workers' paychecks were rising twice as fast as the cost of living, and families were spending the extra dough on cars, TV sets, radios, washers and dryers, clothes, food, and tranquilizers—so many that every U.S. man, woman, and child could have had six pills each in 1956.

The Cold War angst of baby boomers and their parents was covered in a veneer of unprecedented economic prosperity that brought with it not only a cornucopia of consumer goods but also entertainment options. (Why worry about annihilation? Let's pack the car and go to Disneyland! Don't forget the tranquilizers!) Indeed, the amusement park's easy freeway access and ample parking hastened the escape from atomic-bomb tests and civil-defense drills. The fabulous lands of Fantasy, Frontier, Adventure, and Tomorrow quickly became a must for mid-twentieth-century vacationers. Disneyland's turnstiles counted one million visitors in less than two months.

Next to the purely escapist amusements of Davy Crockett's Frontier Stockade and Sleeping Beauty's Castle, an exhibit devoted to the wonders of modern chemistry could have seemed incongruous. But the Hall of Chemistry, sponsored by Monsanto, was as upbeat and gee-whiz as any other Disneyland diversion. Housed in a sleek modernist glass-and-concrete building, the exhibit extolled chemistry's influence in "almost every form of human endeavor" and promoted dozens of Monsanto breakthroughs that would benefit humankind for generations. Visitors learned that Monsanto drugs and disinfectants bettered human health and hygiene, and that its food supplements and insecticides engendered healthier livestock and more abundant harvests. They saw that the company's "miracle fibers" rendered clothing easier to care for and longer lasting, and that plastic's versatility made it the perfect material for applications ranging from rocket nose cones to Christmas decorations. And they marveled at the possibilities ahead, including Monsanto's vision—never fulfilled—that chemicals might economically transform salt water into an unlimited supply of pure freshwater to quench our thirst. (What Monsanto prognosticators failed to foretell, however, was the St. Louis–based company's prominent role in the development of genetically modified foods.)

That the exhibit did not mention potential environmental and health risks from the revolution in chemical applications is not surprising: cautionary tales about the detrimental effects of chemical substances, including Rachel Carson's seminal work *Silent Spring*, were years in the future. The Environmental Protection Agency had yet to be created, and the U.S. government would not begin to attempt regulating chemicals until the 1970s, by which time sixty-two thousand substances were already in U.S. commerce. Besides, mid-twentieth-century America was after something else: reassurance that scientific discoveries in corporate and government labs were leading to a brighter future—even as the Cold War threatened to end it all in a mushroom cloud at any second. In a 1957 poll by the National Association of Science Writers, nearly 90 percent of the U.S. public agreed the world was "better off because of science." It was a heady time for a chemical manufacturer, and Monsanto used its Disneyland exhibit to herald its achievements.

Inside the Hall of Chemistry, which stood at the gateway to Tomorrowland until 1966, man was in control. Visitors learned how thousands of consumer products owed their existence to chemical engineering that transformed natural materials into inexpensive and durable synthetics. To symbolize this technological feat, a giant-size palm holding a ball of fire was painted on the ceiling above eight huge test tubes at the center of the exhibit. This was the hand of chemistry, man exerting power and influence over nature. The exhibit didn't explain any of the complex processes by which chemical transformations occur. All was simply magical, like the rest of Disneyland.

Children of the Synthetic Age, boomers born into an era of mid-century chemical marvels, were the first generation to experience man-made substances as part of the natural order. Kids played with polyethylene hula hoops, ate from colorful melamine dishes, and drank from unbreakable polystyrene cups. They frolicked in clouds of DDT, the insect repellent sprayed from trucks that cruised their neighborhoods. Mothers cooked on Teflon and swaddled the leftovers in Saran Wrap. Countertops were covered in Formica and polyurethane-foam sofas were upholstered in Naugahyde. The most fashionable clothes came in acrylic, polyester, and nylon. Fuel laced with tetraethyl made cars run smoother and quieted engines. Solvents and bleach scrubbed things clean.

Most certainly, scores of synthetics existed prior to World War II. But they proliferated in a period of postwar optimism when almost anything seemed possible. Led by industry giants including DuPont, Dow, Union Carbide, and Monsanto, chemical industry production increased nearly three times the average of all manufacturing industries between 1952 and 1961. U.S. output of plastics alone increased from three hundred million to six billion pounds in the twenty years between 1940 and 1960. Industry and consumers had a common desire to see new synthetics introduced as quickly as possible. During this go-go period, manufacturers competed to develop entirely new substances and improve on old ones. The industry focused on profits, consumers demanded conveniences, and chemicals flooded our world. By the turn of the twenty-first century, more than eighty thousand chemicals were registered with the EPA.

World War II had set the stage for spectacular expansion of synthetics and their applications. The spotlight was on chemical companies, and they performed like stars, producing record quantities of synthetic rubber, aviation fuel, and such specialty metals as magnesium and aluminum. Industry leaders including DuPont, Union Carbide, Harshaw Chemical, and Hooker Electrochemical helped out on the Manhattan Project, the government's top secret atom bomb venture. While DuPont's Teflon would eventually wind up as a nonstick coating for frying pans (and as a descriptive for particularly slick politicians), it had a more auspicious start coating the valves and gaskets in the atom bomb. In case after case, the theaters of World War II put synthetic substances to the ultimate test, and under the harshest conditions, synthetics came through.

Plexiglas—strong, lightweight, transparent, and aerodynamic—proved to be the perfect new material for molding cockpit canopies and gunners' compartments. The giant Douglas B-19 superbomber sported Plexiglas enclosures on its nose, tail assembly, top gunner turrets, and side windows. Oil additives known as Acryloids helped win crucial battles in cold climes. Tanks and artillery with Acryloid additives remained functional in subzero temperatures, giving Allied tanks and artillery the upper hand over the

Germans when their equipment froze up. (The Rohm and Haas researcher who discovered Acryloid crowed, with tongue halfway in cheek, that he was responsible for the Russian victory at the battle of Stalingrad.) The insecticide DDT—inexpensive, easy to apply, and toxic to a broad spectrum of pests—kept troops safe from diseases such as typhus and malaria. DDT came home a war hero, heralded as a miracle pesticide and destined for widespread agricultural and commercial use. Paul Müller, the Swiss chemist who discovered it, was awarded a Nobel Prize in 1948.

Applications for plastic alone seemed endless during the war. Boots, tents, clothing, rafts, parachutes, canteens, and even bugles made from synthetics became trusted battle partners. They didn't rot, rip, melt, or dent, persuading doubters who had frowned on the materials as frivolous or shoddy. In the popular press, plastics were heralded as "materials in their own right," moving from a "lighthearted aesthetic" to "tough functional seriousness," according to *Harper's* in 1942. At a convention the same year, a chemical industry economist and historian predicted that synthetic materials would have "more effect on the lives of our great-grandchildren than Hitler or Mussolini."

Suddenly, synthetics had cachet, and uses aplenty. After the scarcities and sacrifices of war, materialism ran rampant. Rosie the Riveter, her returning war hero, and their growing broods needed new cars and gadgets, not to mention houses and furnishings. Synthetics propelled the production of pink flamingos for the lawn, Tupperware for the kitchen, and fin-tailed automobiles for the garage. In 1953, after a decade of creeping use of vinyl car seats, curvaceous Plexiglas light covers, and plastic convertible tops, Chevrolet dispensed with steel underpinnings and introduced the fiberglass-frame Corvette roadster. Today, each vehicle that rolls off the assembly line contains nearly $2,000 worth of chemical processing and products, including 290 pounds of polymer.

Synthetics were revolutionizing construction, too, with materials ranging from plastic pipes to drywall. In 1957, Monsanto added the House of the Future to its Disneyland oeuvre. The model home, composed of four cantilevered wings suspended aboveground, was constructed entirely of plastic, eschewing conventional interior and exterior building materials. Walls and ceilings sloped and curved and were covered in "wipe-down and wash-off" laminates. Thick Acrilon carpeting and easy-care vinyl covered the floors (cushioned with a vinyl sublayer for extra foot comfort). Eames fiberglass chairs—upgraded with foam coverings—surrounded the Herman Miller family room table. The House of the Future and its furnishings were almost 100 percent man-made, leading Monsanto to boast, "Hardly a natural material appears in anything like its original state anywhere in the building." While plastic houses never caught on, the headlong rush into plastics was unstoppable. Synthetic polymers—chains of small molecules joined together chemically to form one giant molecule—penetrated the production of everything from toys and picnicware to packaging materials and phonograph records. Such was the success of plastic that by 1979, its global production volume surpassed that of steel.

The roots of this modern material world date to nineteenth-century discoveries involving coal. Industrialists were thrilled to learn that coal tar—a thick, smelly, toxic black waste product that results from heating coal to make gas or coke—could be used to derive dyes, aspirin, saccharin, perfumes, the earliest plastics, and even TNT. The study of coal-tar chemicals launched the production of synthetics based on organic chemistry. Since 1856, when a British chemist named William Henry Perkin accidentally discovered the first synthetic dye (it was mauve), organic compounds have proved to be the most varied and pervasive substances in industry. Carbon atoms, the backbone of organic compounds, allow for millions of variations in molecular structure, setting the stage for wide-ranging synthetic innovations.

Most of the chemicals in use today are processed from hydrocarbon-containing raw inputs such as oil and natural gas. In the mid-twentieth century, new, improved, and vastly cheaper methods for obtaining basic chemicals from petroleum and natural gas pushed aside coal tar as the primary source of organic chemical production. Today, the chemical industry is the largest single user of both oil and natural gas in the United States. It uses 7 percent of U.S. petroleum products and 12 percent of U.S. natural gas, and its dependence on these nonrenewable fossil fuels is a huge challenge facing the industry in the twenty-first century.

Inside a petrochemical plant, oil and natural gas are turned into bulk petrochemicals, which are then processed into basic chemical building blocks. In 2004, the U.S. chemical industry produced more than 138 billion pounds of seven bulk petrochemicals: ethylene, propylene, butylenes, benzene, toluene, xylenes, and methane. These are the starting points for tens of thousands of chemical products. Through a series of chemical reactions, these

bulk feedstocks are converted into the chemical intermediates and, ultimately, the finished goods that distinguish
our quality of life.

Inorganic chemicals, which typically do not contain carbon, also play an important role in the creation of chemical synthetics. Derived from metals, metallic compounds
(including salt), and nonmetallic minerals, inorganic
chemicals are often compounded with organic substances.
Some inorganic chemicals, such as hydrochloric and sulfuric acids, are used as reactants in industrial processes,
though they are not incorporated into the final product.
However, certain metals and halogens, such as chlorine
and bromine, are critical to the production of finished
goods.

It would be difficult to find any industry that did not
rely on at least one chlorine-containing chemical. Chlorine
is a key ingredient in making plastic pipes. Its ability to kill
bacteria and viruses made drinking water safe. The tropics
teemed with disease-carrying mosquitoes until chlorine-
based pesticides tamed the swarms. Chlorine-based refrigerants modernized food storage and cooled houses and
cars.

These breakthroughs were based on organochlorines, a
family of compounds produced when chlorine gas reacts
with organic matter. Organochlorines have proved themselves the stable and durable chemical workhorses of the
industrial world. But their persistence also makes them
problematic. Most organochlorines resist the natural
processes of degradation. Once they're introduced in the
environment, organochlorines take decades to break down.
These substances are globe-trotters: they travel north by
hitching rides on wind and water currents and with some

migratory species. As a result, organochlorine pollutants taint even the most remote regions of the Arctic. Organochlorines have another quirk that makes them hard for humans to manage: they are attracted to fatty tissue and magnify up the food chain at higher and higher concentrations by accumulating in the body fat of living organisms. Caribou in Canada's Northwest Territories, for example, have ten times the level of certain organochlorine pollutants than the lichen on which they graze. Wolves that feed on the caribou have sixty times the level as the lichen, and so on. This means that even the small releases of these pollutants can have significant impacts.

Organochlorines dominate the list of persistent organic pollutants (POPs), which scientists have identified as some of the most toxic substances ever synthesized. Decades of scientific studies link persistent organic pollutants to an array of ill effects in animals and humans, including cancer, damage to the central and peripheral nervous systems, reproductive disorders, and disruption of the immune system. Everyone on the planet (including newborns) carries traces of persistent organic pollutants, according to the United Nations Environment Programme, which is why the UN has undertaken an ambitious treaty to rid of the world of twelve of the worst of them. POPs chemicals are pollutants without passports, traveling from tropical and temperate climates before settling in the colder regions of the poles and entering the food chain. Canada's Inuit hunters, a highly exposed population because of their location and way of life, marked the POPs treaty's May 2004 effective date with a feast of whale, seal stew, fish, and caribou. The celebration, however, was premature: so tenacious are these chemicals that it will take generations to rid them from the world, even if we never release another ounce.

All of the "dirty dozen" substances covered by the UN treaty, known as the Stockholm Convention, are organochlorines, including DDT, the erstwhile miracle pesticide, and PCBs, the versatile industrial compound introduced more than fifty years ago and phased out or placed under strict control by most Western countries in the 1970s after scientific studies revealed their potent toxicity. Eight others are aldrin, chlordane, dieldrin, endrin, heptachlor, mirex, toxaphene, and hexachlorobenzene, which is also a highly dangerous by-product of making certain cleaning compounds. Two unintentionally produced families of compounds, dioxins and furans, complete the list. They are released in the production of pesticides, polyvinyl chloride, and chlorinated solvents. That all these substances remain a global threat in the twenty-first century speaks volumes about not only their persistence but also our approach to chemicals: the world embraced synthetic substances as miracles and used them as freely as water without knowing what they could do to the environment and people. Today, we are still struggling to understand the risks from current levels of chemical exposures—even as thousands of new substances are registered with the EPA each year.

This do-it-now-ask-questions-later approach is woven into the fabric of our consumer culture, which places a higher value on innovation than on safety and sustainability. Consider: the latest gadgets from Apple make instant headlines worldwide while it barely registers when the company announces a free computer take-back program that will keep millions of pounds of e-waste out of landfills.

Through the art of chemical synthesis—taking known materials and manipulating them to produce new materials—chemistry created not only advancements but also new human needs. Twentieth-century products viewed as luxuries when introduced—cars, telephones, televisions, refrigerators, washing machines, and personal computers—evolved into essentials. At the same time, the status, convenience, or pleasure conferred by nonessential items made some as highly coveted as basic comforts.

Plastics, a now ubiquitous innovation of the twentieth century, begot plastic (and a famous movie line in the 1967 classic *The Graduate*). Bank credit cards, introduced in the 1950s, enabled consumers to spend beyond their means, spurred on by corporate advertising encouraging the quick replacement and total consumption of material things. Accordingly, consumer indebtedness has skyrocketed, too, reaching $2.38 trillion in the United States in November 2006. Among the 46.2 percent of bank card holders who carry a monthly balance, the average is $5,500, according to the Federal Reserve's most recent triennial Survey of Consumer Finances.

Chemicals underlie the consumer culture that now engulfs the world. According to a 2006 report by the University of California, the amount of chemicals produced or imported by the United States in one day would fill up 623,000 tanker trucks with a capacity of 8,000 gallons each. If placed end to end, these trucks would reach 6,000 miles from San Francisco to Washington, D.C., and back again. So pervasive are synthetics that their smells—sweet and sour like a plastic shower curtain or musky like fertilizer—are as familiar as a whiff of grass. Synthetics shadow us as we move throughout each day. Consider your surroundings at this very moment. Most likely, the furniture was manufactured with solvents, coatings, and glues. Your clothing was spun from man-made fibers, or, in the case of cotton, treated with synthetic pesticides. If you're wearing eyeglasses or contact lenses, they fit comfortably because of plastic. If you've got background music bathing the room, it's because electronic equipment containing lasers, chips, and wiring is decoding a synthetic disk made from metals, plastics, and dyes. Chemicals envelop us—they are integral to virtually all consumer and industrial goods from food to cosmetics, lumber to appliances, fuels to textiles, plastics to electronics. The pervasiveness of man-made substances is so complete that in the space of less than one hundred years, they have become completely ordinary.

What's extraordinary is that we know so little about the risks posed by their inherent toxic properties. Apart from disasters such as the one in Bhopal, India, in 1984, where a chemical plant released the raw ingredient for a pesticide, methyl isocyanate, into the air, causing the immediate deaths of three thousand people and seriously injuring another five hundred thousand, chemical production occurs largely out of public view. Unless something goes terribly wrong, we barely note the activities beneath the smokestacks at more than 13,300 chemical plants around the United States. Furthermore, we trust that regulations are in place to protect people and the environment from the most dangerous chemicals—and the recurrence of past blunders such as the toxic Love Canal neighborhood in Niagara Falls, New York, and the dioxin scare at Times Beach, Missouri. As we shall see, such trust is misplaced when it comes to the chemicals used in everyday products.

In 1976, after much debate and vigorous industry opposition, Congress gave the Environmental Protection
Agency authority over testing of chemicals and pesticides.
By then, the nation—indeed, the world—had learned the
hard way how chemicals that appear benign when first
introduced—PCBs and DDT, for example—can wreak
havoc with the environment and human health. The
EPA's charge was twofold: to assess and manage the risks
of sixty-two thousand chemicals already in commerce
and to review new chemical introductions. This legislation, the Toxic Substances Control Act, aimed to safe-
guard the public interest by obtaining more information
on chemicals and to control those that present "an unreasonable risk of injury." But Congress put regulators
on a short leash: their actions could not "impede unduly
or create unnecessary economic barriers to technological
innovation."

In effect, the toxics legislation we have lived with for
more than thirty years prevents regulators from acting unless they are able to demonstrate how the benefits of
restricting a chemical prevail over the costs of such restrictions to business and society. This hurdle "places too high
of a bar for EPA to jump to assure the health of the public
and protection of the environment," according to Dr.
Lynn Goldman, who was the EPA assistant administrator
in charge of the Office of Prevention, Pesticides and Toxic
Substances from 1993 to 1998. Indeed, the EPA has not
attempted to ban a toxic chemical since 1989. The result is
that existing chemicals "are considered safe until proven
guilty, even when found in breast milk and even as toxicology evidence accumulates," said Goldman, whose assess-

ment is backed up by a series of reports from the Government Accounting Office.

In 2005, the GAO urged Congress to consider beefing up the regulations to give the EPA more authority to assess risks of existing chemicals. Indeed, suggestions that TSCA needs to be strengthened are nearly as old as the legislation itself. Volumes have been penned about the limitations of the toxics act. Besides reports by the GAO, the National Academy of Sciences, the congressional Office of Technology Assessment, Environmental Defense, and the EPA have each identified critical areas where TSCA has fallen short of its objectives. But the act has not undergone any significant legislative action since it was passed more than thirty years ago.

One of the biggest problems with TSCA is that it gives blanket approval to sixty-two thousand chemicals in commerce prior to the legislation's implementation. No questions were asked. No hazard data were required. It's not surprising, therefore, that 99 percent (by volume) of chemicals used today are older substances that were grandfathered in under the toxics act, according to Inform, a New York–based research organization. Chemical manufacturers are responsible for reporting any potential problems they discover about these products. This makes EPA chemical risk management dependent on information *volunteered* by industry, and industry has little incentive to look for damning problems or to disclose all it knows.

When manufacturers make information filings to the EPA, the manufacturers themselves often deem them confidential—as allowed by the toxics act. In 1998, for

example, 40 percent of the substantial risk notices filed
by manufacturers asserted the *identity* of the chemical
was confidential. That same year, more than 65 percent of
the information filings made through TSCA were claimed
as confidential. The EPA has pursued some spurious confidentiality claims, but it does not have resources to put an
end to the practice. Under the toxics laws, information
that manufacturers claim to be confidential "shall generally be treated as such as long as no statute specifically requires disclosure," according to a 2005 GAO report. "It's
ludicrous how far these companies will take confidentiality claims," said Irving "Pep" Fuller, formerly the EPA's
counselor for international affairs and now a chemical industry consultant. "They'll say, 'We manufacture chemical
blank.'" The EPA has turned to computer modeling—
predicting the toxicity of a chemical based on its structural
similarity to chemicals with known effects—to glean in-
sights. Says Fuller, "EPA staff have had to become Sher-
lock Holmes to try to postulate what might happen." As
for chemicals introduced after TSCA's implementation,
they are "barely assessed," said Goldman, a pediatrician
who is now a professor of occupational and environmental
health in the Johns Hopkins Bloomberg School of Public
Health.

   According to the GAO, the problem stems from toxics
regulations that allow the EPA access to little, if any, test
data. A manufacturer's notice to the EPA that it intends to
begin manufacturing or importing a chemical (of which
the EPA receives between fifteen hundred and three thousand a year) is not required to contain toxicity or exposure
information, and few of them do. Such data, however, is

essential for the EPA to evaluate the health and environmental threats of any chemical.

Since the implementation of TSCA, the EPA has attempted outright bans on only two chemicals. The toxics legislation itself set out the regulations for one, PCBs. Widely used in twentieth-century industrial and commercial applications such as electrical transformers, fluids and lubricants, inks and dyes, adhesives, and protective coatings, PCBs were phased out because of their potent toxicity and links to serious illnesses such as cancer. Their persistence and stability, traits that made them attractive to industry, allow PCBs to migrate thousands of miles from their original source and linger for decades. These persistent organic pollutants can be found in Arctic polar bears and are still being detected in the U.S. population nearly thirty years after they were banned.

In comparison to the fate of asbestos regulation, however, the EPA's restrictions on PCBs are a wild success. In fact, the agency's experience attempting to regulate the fibrous mineral compound shows how impossible it is to restrict toxic substances through regulatory means. After years of research, public meetings, and regulatory impact analyses, the EPA in 1989 exercised its authority under TSCA and prohibited the future manufacture, importation, processing, and distribution of asbestos in almost all products. A legal challenge by industry, however, resulted in a landmark ruling by a court of appeals that undid much of the EPA's ban and all but eliminated the agency's ability to use the toxics act to restrict problem chemicals.

In addition to bans on PCBs and asbestos, the EPA has restricted only three other chemicals in existence prior to the toxics act: dioxin, the by-product of certain industrial processes that's thought to be one of the most toxic chemicals ever made by humans; hexavalent chromium, a dangerous form of the metal used in metal finishing, chromium chemical production, paint pigments, leather tanning, and some wood preservatives; and chlorofluorocarbons (CFCs), ozone-eating chemicals commonly used as coolants for home and car air conditioners and in the making of fast-food containers. DDT was banned in 1972, prior to the enactment of TSCA. These chemicals are the devils we know. As for other contaminants now emerging in the scientific literature as substances of concern, the EPA cannot act swiftly or decisively to weed out the bad actors. While the EPA's Office of Pollution Prevention and Toxics (OPPT), which administers TSCA, cannot change the law itself, critics contend the culture of OPPT is too closely tied to industry. As a result, its managers are unable to see where improvements in chemical assessment and management programs could be made, even within the existing policy framework. Citing their frustration and disappointment with OPPT, three members of the EPA's National Pollution Prevention and Toxics Advisory Committee (NPPTAC) resigned in October 2006. Among them was Joe Guth, the legal director of the Science and Environmental Health Network, who told me OPPT "prefers not to even consider change."

When questions arise about chemicals, manufacturers can—and often do—use the absence of information to argue that a substance is harmless. Under our regulatory structure, ignorance is rewarded: manufacturers have no obligation to test for the safety of the substances they sell.

Indeed, they have a financial incentive not to do so. Meanwhile, the EPA has all but given up on trying to use the toxics act to better understand the potential hazards of tens of thousands of chemicals. The agency lacks the statutory power to request data on a chemical prior to proving it causes harm. And it can't make that kind of risk calculation without the data it is seeking. This bureaucratic Catch-22 has created a gaping void in information about the potential hazards and long-term effects of our dependence on chemicals.

Reform of the nation's toothless toxics regulations unfailingly is met with resistance by the chemical industry. More testing and greater regulatory control would cost billions of dollars and practically return society to the Dark Ages, goes the argument. What the chemical industry fails to mention, however, is that food and drug companies have managed to thrive under regulations that give authorities much greater latitude in testing and reviewing products.

The European Union is already marching ahead with tougher toxics laws that require manufacturers to produce risk and exposure information on some thirty thousand chemicals. "We need to shift the responsibility for chemical safety to the chemicals industry itself," said Margot Wallström, European Commission vice president. "Taking responsibility for your own products is standard practice everywhere . . . Why should the chemicals industry be treated in a different way?" The new policy, known as REACH (Registration, Evaluation, Authorisation and Restriction of Chemical Substances), promises to close the knowledge gap on chemicals, reverse the burden of proof so that manufacturers must demonstrate a product is safe before it gets to market, and make more information available to consumers.

The chemical industry—both in the European Union and the United States—found much to criticize in REACH. It even enlisted the Bush administration's help, and the administration launched a strong attack on the REACH proposal, calling it "costly, burdensome and complex." The American Chemistry Council, the industry's largest U.S. trade organization, proudly described how it "rallied opposition to the draft proposal, including a major intervention by the U.S. government . . . These efforts helped to build an aggressive position worldwide, and brought about significant concessions."

Greg Lebedev, former ACC chairman, explained how it worked in a 2004 speech: "We arranged for multiple elements of our government—the Department of Commerce, the U.S. Trade Representative, the Environmental Protection Agency, and the Department of State—all to express the understandable reservations about this proposed rule and its trans-Atlantic implications. I only wish that we could exert so much influence every day." Special-interest lobbying is not unique to the chemical industry. In this case, however, it appears that industry had unfettered access to federal policy makers, and the public was largely shut out of all discussions regarding REACH, House Committee on Government Reform investigators found.

When threatened, the chemical industry has historically used size, strength, and spin to squash detractors, waging all-out battles against those who would put public health ahead of chemical proliferation. Industry leaders, confident that their products were always in the public interest, were dismayed when others disagreed, and they strongly suggested the doubters must be cowardly, misinformed, or both. At a national conference on tetraethyl lead in 1925, Frank Howard of the Ethyl Gasoline Corp. described the discovery of leaded gasoline as a "gift of God." As Howard explained, "I think it would be an unheard-of blunder if we should abandon a thing of this kind merely because of our fears." Certainly, the chemical giants of the twentieth century did their best to encourage consumers to have faith in the latest chemical breakthroughs and to place their trust in the scientists who made them possible.

This trust-us-we're-experts approach was displayed throughout much of the twentieth century, when chemical manufacturers were prominent exhibitors at expositions and world's fairs designed to show off industry's latest advancements and build confidence in the economy. At Chicago's Century of Progress Exposition in 1933, on the heels of the Great Depression, General Electric put "electricity wizardry" on display inside an exhibit called the House of Magic. Those who listened to the *GE Symphony Hour* on the radio already knew the company called its research laboratory by that name. And despite the protests of some GE scientists, the House of Magic name stuck, spurred by the enthusiastic endorsement of the advertising department, which had learned that radio listeners— mainly women—were taken by the notion of magic and miracles coming from the company's research lab.

During the same era, DuPont adopted the slogan "Better Things for Better Living . . . Through Chemistry." And who could argue? At the 1936 Texas Centennial Exposition, more than 1.5 million visitors to the Wonder World of Chemistry learned "how DuPont chemists take Nature's raw materials and convert them into articles we all know and enjoy today." Displays of plastics, artificial rubber, antifreeze, cleaning solutions, refrigerants, insecticides, dyes, and paints highlighted the exhibit. University students hired as lecturers and demonstrators reported to company officials that most visitors felt good about these miracle materials and the company behind them. One woman found it "wonderful how DuPont is improving on nature." Meanwhile, the company's sponsorship of the *Cavalcade of America* radio show familiarized millions of Americans with the "Better Living" slogan, which was the brainchild of one of the country's top ad men.

By the time the 1939 New York World's Fair opened, exhibitors such as AT&T, General Electric, DuPont, Standard Oil, General Motors, Ford, Chrysler, and U.S. Steel were wowing visitors with all kinds of industrial sorcery. Visitors could chat with robots at the AT&T and Westinghouse exhibits and pose questions to a "Magic Car"—it answered by opening and closing its doors and blinking its lights—at the Chrysler pavilion. At the DuPont exhibit, Miss Chemistry, fashionably clothed in garb made from DuPont's synthetic fibers, appeared from an empty test tube, "as if she herself, in addition to her costume and accessories, had been created by chemistry."

At the 1964 New York World's Fair, attended by some fifty-one million visitors and perhaps the biggest-ever PR bonanza for science-based companies, one of the most popular exhibits was GE's Carousel of Progress featuring Walt Disney's Audio-Animatronics, a form of robotics that used lifelike mechanical figures controlled by computers. The audience revolved around a stage divided into slices of progress and watched all-American robots enjoying the fruits of consumer culture—electric lights, phonographs, refrigerators—at different moments in the past century.

Those who questioned the wisdom of this headlong rush into a world built on a foundation of synthetics drew bull's-eyes on their backs, as Rachel Carson learned with the publication of *Silent Spring* in 1962. Carson, a respected biologist and best-selling author, had a "remarkable knack for taking dull scientific facts and translating them into poetical and lyrical prose that enchanted the lay public," according to *The New York Times*. In what would be her final book, she wrote passionately and persuasively that indiscriminate use of pesticides, including DDT, was harmful to people, birds, fish, livestock, crops, and flowers, and warned about the long-term effects of such misuse. Carson, calm, soft-spoken, and meticulous in her work, was attacked by industry and labeled an alarmist and a "hysterical woman" who was unqualified to write such a book. Leading the well-organized charge was Monsanto, sponsor of the Hall of Chemistry at Disneyland. In the October 1, 1962, issue of *Monsanto Magazine*, Carson's work was parodied in an essay titled "The Desolate Year," which described what the world would be like if pesticides were nonexistent for one year. Charlie Sommer, Monsanto's president at the time, explained, "We felt we had a responsibility to demonstrate the other side of the tradeoff—namely, the perils of abandoning pesticides." Readers, however, understood that Carson was not advocating that we leave pesticides behind but rather that we take precautions with their use. The message in *Silent Spring* helped launch the modern environmental movement. That pesticides and other widely used synthetic chemicals now are present in everyone in the world would not have surprised Carson. She recognized the delicate, inextricable link between humans and the environment. That industry downplays the significance of such findings and resists efforts that would help us learn more about chemicals would not have surprised her, either. Nor should it surprise any of us. Inexcusable, however, is how Congress—through its failure to put muscle behind our toxics laws even as our understanding of chemical hazards grows—has made every one of us a test animal in a vast, uncontrolled experiment. As I describe in the following chapters, even today, those who speak up are, like Carson, met with derision by a chemical industry striving to maintain a U.S. regulatory system that can only shrug at this travesty.

KERMIT'S BLUES: ATRAZINE AND FROGS

In 1998, when Tyrone Hayes accepted research funding from the makers of atrazine, the most heavily used agricultural weed killer in the nation, the compound was under special review by the Environmental Protection Agency. A decade or so earlier, the discovery by the U.S. Geological Survey that atrazine runoff from farmers' fields was fouling the nation's water systems threw the safety of the herbicide into question. So the EPA in 1991 used its authority under the federal Safe Drinking Water Act to lower atrazine drinking water standards to no more than 3 micrograms per liter of water, or 3 parts per billion. At that microscopic level, regulators believed they had created a safety margin that would protect even the most vulnerable from any possible harm. More research piled up on atrazine, and it wasn't particularly reassuring. A rat study suggested atrazine might cause mammary tumors. That helped trigger a special EPA review of atrazine and related weed killers known as triazines. Atrazine was also classified as a possible human carcinogen. Clearly, regulators needed to know more about what the weed killer was doing to people and the environment. So the EPA did what pesticide regulations require it to do when seeking information about the hazards of a substance in widespread use for several decades: it asked the manufacturer to provide it. The Federal Insecticide, Fungicide, and Rodenticide Act places the costs of research and testing on companies that profit from a substance—not on U.S. taxpayers.

Novartis, which became Syngenta through a 2000 corporate merger, is the principal manufacturer of atrazine. The company hired Tyrone Hayes, a professor of integrative biology at the University of California at Berkeley, to join an expert panel convened to study atrazine's ecological effects. He had all the right credentials. Hayes, a frog expert, patented an amphibian assay to examine endocrine disruptors, or substances that interfere with natural hormones. He holds a biology degree from Harvard University and a doctorate in amphibian development from Berkeley, where he was tenured at age thirty and later became the university's youngest full professor. Now nearing the age of forty, Hayes is science's version of a rock star. No matter the time of year, he pads around the Berkeley campus in flip-flops, nylon shorts, and a sweatshirt, his bushy hair often pulled into a ponytail and his beard, when he has one, finished into rakish little points. A shade taller than five feet with a robust chest and muscular legs, Hayes's squarish, compact body makes him look like a coach's dream halfback. Wherever he goes, he's surrounded by students—his own and others'. At a national toxicology conference, I watched Hayes, dressed in a black suit (his favored attire for all speaking engagements), autograph programs at the request of bedazzled young scientists. Not only is he a prolific researcher who has published dozens of peer-reviewed papers, he's a riveting speaker with a rare gift among his brethren: Hayes can explain his work to anybody, including those who have never cracked a biology textbook. And he's unfailingly good-natured about it.

The eldest son of a carpet layer and a homemaker, Hayes was born and raised in Columbia, South Carolina. As a kid, he hunted for amphibians near his family's house. He was intrigued by their transformation from egg to tadpole to frog. As a scientist, he still finds the transparency of amphibian development riveting. "What is fascinating about the animal is [that we can] watch developmental events, including . . . the metamorphosis process," Hayes said. Amphibians don't have shells, membranes, yolk sacs, or placentas, so they're terrific tools for learning about the developmental effects of various substances on the endocrine system, which is the primary focus of Hayes's research and teaching. Consider, during the transformation from tadpole to frog, all the genes that make a tadpole turn off, and the genes that make a frog turn on. "The genes are turned on and off by hormones that are chemically the exact same hormones that circulate and regulate [human] physiology," Hayes said. In other words, as Hayes put it, "Hormones are hormones are hormones." That is, they work the same way across vertebrates—from frogs to mammals. Just as hormones are the reason tadpoles turn into frogs, they are in humans the reason we develop in the womb, grow into adults, turn food into fuel, make sperm or eggs, and sprout beards or breasts.

Hormones, released by endocrine glands directly into the bloodstream, are the chemical messengers of the endocrine system. Once released, they travel to distant target cells, triggering the signals to carry out certain functions. Given the endocrine system's powerful influence over health and behavior, it's not hard to understand why substances that cause glitches in this intricate physiological scheme raise serious concerns. Environment Canada, the equivalent of the EPA, warns, "Even very subtle effects on the endocrine system can result in changes in growth, development, reproduction or behaviour that can affect the organism itself, or the next generation." That's why scientists such as Hayes are on the lookout for substances that might mimic the action of a naturally produced hormone, block receptors in cells receiving hormones, or alter the concentrations of natural hormones.

Accompanied by students from his lab, Hayes has covered thousands of miles throughout the United States and Africa collecting frogs and conducting fieldwork for his studies. When he sees a substance interfering with hormones in frogs, it suggests to him that human effects are also likely. And when he reads studies showing that Missouri farmers with poor semen quality are excreting urine with atrazine levels high enough to chemically castrate frogs in his laboratory, Hayes says his worries grow frighteningly real. "Many of us believe that frogs are modern-day canaries," Hayes said. "Maybe what they're telling us is that it's just a matter of time."

Hayes is especially good at teaching his students the importance of the endocrine system and how hormones re-

late to every aspect of living. His knack for turning real-
life situations into teachable moments helped earn him
Berkeley's Distinguished Teaching Award in 2002. He fa-
mously uses fetal ultrasound images of his two children,
Tyler and Kassina (his daughter is named after an African
frog), when he lectures on human development. He re-
worked his syllabus after some female undergraduates
asked him why aspirin relieves their monthly cramps and
he realized menstruation is barely discussed in most en-
docrinology courses. And he encourages debates over the
ethics of such controversial issues as the appropriate uses
of growth hormones. "The fact that he's so open and so
approachable makes his classes the best I've ever taken in
biology," said Victoria Ngo, a student who was inspired
enough to seek a job in Hayes's amphibian lab.

When it came time for his own college education,
Hayes filled out an application for just one—Harvard University—mainly because of the TV sitcom *Green Acres*. "It
sounds stupid, but I applied because Oliver Wendall Douglas [a country-lovin' blue blood played by the late Eddie
Albert] went there, so I figured it had to be good," he
said. He took his inaugural plane ride when he left home
bound for Cambridge, the first in his family to attend college. It wouldn't be stretching it to say Hayes was as in-
genuous about the world of expert scientific panels when
he signed on with Syngenta as he had been about acade-
mia. Luckily, at Harvard he found a perfect fit for his interests and skills in a laboratory run by amphibian expert
Bruce Waldman. And he met his wife, Katherine Kim, as
an undergraduate. They married as soon as Hayes gradu-
ated. His foray into the world of corporate-sponsored science, however, would not prove so agreeable.

By the time Hayes's lab began studying atrazine, the herbicide had been shown to produce endocrine-disrupting effects in a variety of animal studies. It was of special concern to regulators because of its persistence in the environment. Depending on conditions, it takes atrazine from 41 to 237 days to break down in water, according to the EPA's environmental risk assessment on the weed killer. In cold-water environments such as Lake Michigan, the agency noted that atrazine can persist for years. The general population is exposed to this weed killer through food, water, rain, and dust, and from its use on lawns and crops. You don't have to be a farmer or live near one to be exposed: atrazine and its degradates—the substances formed as the pesticide reacts chemically, photochemically, or biologically in the environment—are commonly found far from usage areas because of the way the weed killer hitches rides in the wind and on water. In a 2003 paper published in *Environmental Health Studies*, two EPA scientists wrote that nearly 60 percent of the U.S. population is exposed daily to the weed killer.

Because of its popularity and wide use, the CDC decided to measure the levels of atrazine in people as part of its ongoing reports on human exposures to environmental chemicals. CDC scientists based their study on the notion that the human body most commonly breaks down atrazine into a metabolite called atrazine mercapturate before it is excreted in urine. In all three national reports on human exposure to environmental chemicals released to date, urinary levels of atrazine mercapturate were below the limit of detection. But Dana Barr, the chief of the CDC's pesticide lab, cautions this is not the end of the story. "We now understand it's very important to measure other atrazine metabolites," she said. The reason: the CDC has discovered that several other human metabolites of atrazine are critical biomarkers of the herbicide, and published these findings in 2007. As a result, Barr said, "Atrazine exposures in the general population have likely been underestimated." In the future, the CDC will include measurements of about a dozen atrazine metabolites in its national exposure reports, hoping to paint a more accurate picture of human exposure.

Indisputable, however, is that because of the herbicide's wide use in the United States, there is virtually no atrazine-free environment for wild frogs. So Hayes's work for Syngenta would focus on how atrazine affected the hormones controlling frog development. Hayes's laboratory work started simply enough. He added trace amounts of atrazine to some of the water tanks in which he raised African clawed frogs, a common laboratory animal. Then he compared those frogs with other African clawed frogs raised without atrazine in their tanks. The findings were provocative. Male frogs exposed to as little as 1 part per billion, or one-third the level allowed in U.S. drinking water, had impaired laryngeal growth—a bit of a problem for frogs, who rely on basso profundo croaks to attract mates. Hayes also noted what he described as "ambiguous gonads" that were suggestive of hermaphroditism.

To flesh out his findings, Hayes wanted to determine if other types of frogs are similarly sensitive to atrazine. Syngenta, he said, delayed his study requests and wanted to limit the scope of his work to repeating his earlier study. Acting independently, Hayes and his students raised leopard frogs, a species native to the United States, in atrazine-exposed water to see how the herbicide affected their metamorphosis from tadpole to adult. Taking to the field, the Hayes researchers hit the road in a refrigerated truck, collecting eight hundred young leopard frogs at eight field sites stretching from Utah to Iowa. The team collected the animals from a variety of habitats, including golf courses, wildlife management areas, rivers, and runoff from corn-fields. Atrazine levels were measured from each site, and ten thousand gallons of frozen water were taken back to the lab for analysis and further experiments.

Hayes and his team then dissected both the lab-raised and wild animals exposed to atrazine. They found wide-spread hermaphroditism in lab-raised males exposed to 0.1 part per billion, an astonishingly small amount. Incredibly, a measurement of atrazine one-thousandth the size of a grain of salt added to the water was all that was necessary to scramble the sex of the lab-raised specimens. Meanwhile, similarly exposed male leopard frogs collected from the field had reproductive abnormalities comparable to those observed in the frogs from the lab. As Hayes explains, it's clear that wild frogs are exposed to all sorts of herbicides, insecticides, and fungicides. "So we asked: Can we blame atrazine?" They answered that question by dosing frogs with a combination of substances. "We found that atrazine was associated with hermaphroditism 100 percent of the time." In fact, through his fieldwork, Hayes determined there was both a spatial and a temporal correlation between atrazine and hermaphroditic frogs. In other words, "Where we found atrazine we found her-maphrodites, and when we found atrazine we found hermaphrodites," he said. The finding was especially exciting to Hayes because it suggested that by eliminating atrazine, tadpoles at a once-contaminated site could grow up to be normal.

Hayes also came up with a theory about what was causing the abnormalities: trace levels of atrazine exposures—the equivalent of measurements commonly found in the environment—stimulated an enzyme called aromatase, which converts the male hormone testosterone into the female hormone estrogen. Normally, aromatase is silent in males, but in atrazine-exposed animals, estrogen sparked by aromatase induces the growth of female characteristics such as ovaries, eggs, and yolk. Whenever I've heard Hayes discuss these findings at public presentations, this is the point where men in the audience shift uncomfortably in their seats. "The gonads, instead of sperm, have eggs," Hayes explains. "And the frogs are chemically castrated because they do not make testosterone."

From a research standpoint, Hayes's discoveries added up to a holy-shit moment. The findings, though preliminary, implicated atrazine in two of the most debated issues in science: the causes of a shrinking amphibian population and the effects of chemicals that disrupt the intricate functions of hormones. Not only did the studies suggest that atrazine was a likely factor in the decline of amphibians worldwide, they also raised disturbing questions about how atrazine, at background exposure levels, impacts animals and the environment.

From a business standpoint, Hayes's work was bound to attract lots of attention, too. The professor's data suggested that a sentinel species may be sexually impaired by concentrations of atrazine commonly found in the environment. It was just the kind of information that could threaten continued, unrestricted use of the herbicide. Scientifically speaking, Hayes had opened up an exciting path for research. Financially speaking, Syngenta had a vested interest in blocking it.

Syngenta, which earned $1.1 billion in 2007 on sales of $9.2 billion, has a lot riding on the use of atrazine in the United States. That's because the European Union, which has identified atrazine as an endocrine disruptor, banned it in 2005, with exceptions for limited uses where alternatives are not available. The EU restrictions were instituted because the European Commission was concerned about the presence of any toxic contaminant in groundwater at levels exceeding 0.1 ppb (coincidentally, exactly the level of atrazine exposure at which Hayes found sexually scrambled frogs).

U.S. farmers use some seventy-six million pounds of atrazine a year, which makes the herbicide among Syngenta's top-five-selling products, according to company spokeswoman Sherrie Ford. About three-fourths of U.S. acreage dedicated to corn and sorghum is treated with atrazine. So is 90 percent of sugarcane acreage. It's also used on a portion of winter wheat, guava, and macadamia nut crops. And it can be applied at Christmas tree farms and on residential lawns and golf courses, primarily in the southeastern United States, where St. Augustine and Bermuda grass are prevalent. Unquestionably, though, atrazine use is heaviest in the Corn Belt, where every spring farmers fire up their tractors, plant their crop, and spray their fields with it. Farmers cotton to atrazine because it's cheap, easy to use, and effective. At a cost of

about $3 an acre, atrazine kills weeds by blocking their
ability to carry out photosynthesis. While the weeds starve
to death because they cannot convert light and water into
food, crops go unmolested because they are able to detox-
ify the herbicide. As Syngenta toxicologist Tim Pastoor
explained to *The Washington Post*, "[Atrazine] works and
it's inexpensive and that's what farmers love. It's magic for
them. It's like the aspirin of crop protection."

Atrazine was first registered for use in 1958. Chemists at
J. R. Geigy of Switzerland synthesized the chemical in the
mid-1950s. They hoped atrazine would be as effective at
killing weeds as DDT (another Geigy-patented chemical)
was at killing insects. Atrazine quickly became a farmer
favorite, and sales skyrocketed. Between 1959 and 1969,
Geigy Chemical Corporation, the company's U.S. sub-
sidiary, earned $231 million from atrazine and a much less
popular triazine herbicide called simazine. In 1970, Geigy
merged with Ciba to form Ciba-Geigy, and in 1996 Ciba-
Geigy merged with Sandoz to form a new entity called
Novartis, the company that first made contact with Hayes.
Regardless of the name controlling atrazine, the weed
killer has been consistently profitable to its corporate parent, even though its patent expired years ago.

For the first twenty-five years of its commercial life,
atrazine created little fuss. But by the 1980s, it was clear
to researchers that the chemical properties that made
atrazine of such benefit to agriculture—its persistence in
soil and insolubility in water—also made it susceptible to
runoff, leaching, and vaporization into the atmosphere. A
1991 USGS study of eight sites on the Missouri, Ohio, and
Mississippi rivers and on three smaller tributaries found
that atrazine exceeded drinking water standards in 27 per-

cent of the samples and at six of the eight sampling sites. Other government studies confirmed widespread contamination of water resources—especially in the Corn Belt. These contaminant levels, coupled with concerns about atrazine's carcinogenicity, sounded an alarm that led to changes in how the weed killer could be used. In the early 1990s, manufacturers rewrote directions on product labels, instructing farmers to cut back on the amount of applications per acre. Additional distance was required when applying atrazine near streams, lakes, and reservoirs. And atrazine was deleted from the list of acceptable herbicides for all non-cropland.

Despite more than a dozen of such voluntary risk reduction measures, the overall use of atrazine has not declined and the pesticide continues to foul the nation's water resources. A 2006 USGS report—the most comprehensive national-scale analysis of pesticide detections and computer modeling ever compiled—predicted more than one out of twenty streams in the nation's Corn Belt have average concentrations greater than the human health benchmark of 3 parts per billion. Strikingly, atrazine detections led the list of eleven suspected endocrine-disrupting pesticides sampled in the study of agricultural and urban streams. Atrazine levels follow agricultural usage patterns, spiking as high as 50 parts per billion in the spring when rains and irrigation flush the weed killer from newly treated farmers' fields into streams. These spikes also coincide with the breeding season for frogs. Moreover, monitoring by Syngenta of forty U.S. watersheds considered statistically representative of 1,172 others that are potentially vulnerable to atrazine exposure indicated the herbicide has entered streams and rivers at levels that could harm some ecosystems. As of early 2007, two Missouri monitoring sites exceeded the EPA's "level of concern" for atrazine because concentrations reached 50 parts per billion for days at a time. The agency believes atrazine can negatively impact aquatic systems at prolonged exposures of 10 parts per billion.

Hayes's work demonstrates that atrazine is biologically active at exposures as low as 0.1 part per billion, a level the USGS data show is always present in many sampled waterways. "If the EPA accepted my data, atrazine would be off the market," Hayes said. The reason, he explained, is that it isn't simply a matter of the EPA lowering application standards for atrazine. "If you use it at all, there's just no way to get atrazine below 0.1 ppb."

In 2000, studies supported by Syngenta convinced the EPA that the mechanism by which atrazine causes cancer in rats probably does not occur in people. The EPA backed off on its earlier assessment of atrazine as a human cancer threat, clearing a major hurdle toward reregistration. Looming, though, were the questions raised by Hayes's work on low-dose effects in amphibians. This is the point at which the relationship between Hayes and Syngenta went south. Hayes claims Syngenta and the panel tried to "bury" his data and stall additional research. He also claims he was offered substantial money—$2 million—to continue his research in a setting where the panel and Syngenta could control the data; Syngenta denies that such an offer was ever made. Undisputed, however, is that Hayes's association with the panel came to an abrupt, unhappy ending. He resigned in November 2000, citing in his letter a fear that to continue his involvement would hurt his professional reputation. Using funding from various foundations, he pressed ahead with his atrazine research. In 2002, just months before the EPA was to decide whether to reregister atrazine for continued use, two prestigious, peer-reviewed scientific journals—*Nature* and the *Proceedings of the National Academy of Sciences*—published Hayes's findings documenting hermaphroditism in frogs exposed to extremely low levels of the weed killer. In his *Nature* paper, Hayes noted that separate studies on fish, reptiles, and mammals indicated that atrazine triggered adverse effects on sex differentiation in those species, too. And in his *National Academy* paper, he argued that the likelihood is extremely high that wild amphibians are exposed to atrazine levels that cause negative effects because of the prevalence of the pesticide in the environment.

Syngenta and the expert panel wasted no time blasting Hayes's work. In a press statement rife with language more suitable for a courtroom than academe, they announced that three separate studies conducted by Hayes's former panel colleagues could not "replicate the results of earlier, widely-publicized research which alleged that the herbicide atrazine may affect the larynx and sexual development of African clawed frogs." At the time, none of the studies had been published or peer-reviewed. Nonetheless, Ronald Kendall, director of the Institute of Environmental and Human Health at Texas Tech University and chairman of the Syngenta-funded scientific panel, declared, "No conclusions can be drawn at this time on atrazine and its purported effect on frogs."

Science, of course, thrives on open and free debate. It's

how errors are corrected, new theories are adopted, and
old ones are revised. When study results cannot be repli-
cated, it's a sign to researchers that perhaps the original
theory wasn't so good. In that case, the old theory is usually set aside and new hypotheses tested. But always,
among scientists, it's a given that reports will be made
in good faith. The way Hayes tells it, the statements by
Syngenta and the panel members were inaccurate repre-
sentations of their study results. "It was science by press
release," Hayes said. Once the results were made public,
Hayes discovered the data were not in disagreement with
his laboratory's peer-reviewed, published data. Rather, he
said, some of the panel members' experiments were so
flawed by contaminated controls, high mortality, and in-
appropriate measurements of hormone levels that it pre-
vented meaningful comparisons with his work.

"The goal was to create confusion and doubt," said
Hayes—at the very moment when the EPA was evaluating
all available data on atrazine. It's a grave charge, but one,
he said, that is backed up by an analysis of the studies on
frog effects. A report Hayes published in *BioScience* exam-
ines what he described as "the path associated with negative findings" in sixteen experiments testing the effects of
atrazine on the development of amphibian gonads. Hayes
divided the experiment reports into two groups: studies
that reported no effects from atrazine and those that reported effects on the gonads of exposed amphibians. He
then considered five factors—species, study type, study
design, principal authors, and financial sponsorship—that
may have been a significant predictor of outcome. Sure
enough, a pattern emerged: if the study was funded by
Syngenta and authored by company-sponsored scientists,

the results always were reported as showing no negative effects from atrazine on amphibian gonads.

Pete Myers, whose *Our Stolen Future* introduced the concept of endocrine disruption to general readers, argues that Syngenta and its intermediaries borrowed from the playbook of tobacco companies in their dealings with Hayes and the EPA. For years, big tobacco disputed studies that suggested tobacco and secondhand smoke were hazardous and sought to trivialize or discredit studies that raised red flags. Tobacco interests came up with the catchy term "junk science" and affixed it to findings that hurt their cause, while trumpeting as "sound science" findings they liked. The labels that make good sound bites and create plenty of confusion continue to be invoked—usually by business interests—in policy discussions as diverse as global warming, beef imports, air pollution, and genetically modified crops. However, Donald Kennedy, the former head of the FDA and now the editor in chief of *Science*, told Knight Ridder News Service the phrases have more to do with lobbying than lab results. Sound science, Kennedy explained, is strictly in the eye of a beholder. What it boils down to is this: "My science is sound science and the science of my enemies is junk science." It's little wonder that Hayes was labeled a junk scientist by Steven Milloy, a FOX News commentator, former tobacco lobbyist, and publisher of the website junkscience.com, who opined that the professor's work was "more akin to a Brothers Grimm fairy tale than science." Milloy concludes he would "rather kiss a frog" than trust Hayes's findings. Syngenta's home page on atrazine provides links to Milloy's columns on Hayes's work.

As Syngenta postured, the EPA forged ahead collecting data. Tom Steeger, a senior biologist in the agency's Environmental Fates and Effects Division, was in charge of writing a report, issued in May 2003, on the significance of the amphibian findings. Steeger had the onerous task of sorting through data generated by scientists who were openly feuding. He contacted Hayes in April 2002 after Hayes published his findings on atrazine's demasculinizing effects in the *Proceedings of the National Academy of Sciences*. The EPA scientist requested Hayes's raw test data and information about his laboratory methodologies. In a series of e-mails Hayes shared with me (and first made public when the Natural Resources Defense Council and the Environmental News Service requested them under the Freedom of Information Act), Hayes told Steeger that in June 2000 he gave Syngenta his raw data from his initial atrazine study, done under contract with the company, showing laryngeal effects in frogs exposed to 1 ppb of the weed killer. Under federal law, Syngenta is required to report all adverse effects to the EPA. But it had not reported Hayes's findings. Spurred by Hayes's revelation and complaints by the Natural Resources Defense Council, the EPA's Office of Enforcement and Compliance Assurance (OECA) began investigating Syngenta for a possible violation of disclosure requirements. However, Steeger, who declined requests to be interviewed for this book, cautioned Hayes not to expect too much from the investigation.

"Apparently, the Office of Enforcement and Compliance Assurance had its budget curtailed by the Bush Ad-

ministration and OECA's ability to pursue compliance issues has been greatly diminished," Steeger wrote on May 6,
2002. "Your emails outlining when Syngenta (Novartis)
was aware of adverse effect data have been forwarded . . .
What OECA does with the information is uncertain."
(The EPA investigation concluded the evidence did not
warrant enforcement action, spokeswoman Enesta Jones
told me via e-mail. I filed a Freedom of Information
Act request to examine documents pertaining to the investigation. The agency declined, citing exemptions from
the disclosure law including records compiled for law enforcement purposes.)

The controversy over the conflicting atrazine studies got
hotter during the summer of 2002. Television stations,
scientific magazines, and even Hayes's local newspaper
wanted to document the head butting between Hayes and
his former research sponsor. Days before Steeger and
other EPA scientists visited Hayes's Berkeley lab, *The
Oakland Tribune* published a long article with the headline: "Research on the Effects of a Weedkiller on Frogs
Pits Hip Berkeley Professor Against Agribusiness Con-
glomerate." Hayes hated how his laboratory was being
cast as a battlefield. By the end of July, he was fed up with
inquiries from the press asking why Syngenta's studies
could not repeat his findings. Hayes, in an e-mail, con-
fided to Steeger that he was unsure about continuing his
involvement with the EPA review. Steeger urged Hayes to
stay the course.

I am sorry to read that your frustration with the "process"
may preclude your further involvement . . . While I un-

derstand your irritation, frustration, vexation, chagrin, disgust . . . you have brushed up against a corporate reality and your data are perceived as a threat; however, I would strongly encourage you not to succumb to corporate pressure . . . You are correct that your data (not you) are subject to considerable scrutiny since they could potentially impact the regulation of the most commonly used herbicide in the U.S. Your data suggest that a sentinel species may be significantly impacted by environmentally relevant concentrations. It would be very naïve to think that your data would not be subject to scrutiny and calumny especially by those who may be affected financially . . .

Your research . . . will generate controversy because of the chemical, its concentration, and its effects. It would be very difficult to extricate yourself from the public's eye given that atrazine has been detected in most drinking water supplies and because the chemical is near ubiquitous in the U.S. Short of not publishing or using a nom de plume, your audience will soon be much larger and the press will actively pursue you. This is a reality you will not likely escape . . .

If EPA has confidence in your data, we (I) will defend it before the registrant, the Atrazine Panel, the Special Review and Reregistration Division (risk managers), and I will present it to the Science Advisory Panel. You have an opportunity to make an enormous contribution on many levels and not confine your mentoring to students who simply wish to work within the confines of a university. If you believe in your work and you want to impart a sense of right to your students, then I strongly encourage you to work with the EPA in helping to document the effects of atrazine. I can't guarantee that risk managers will act on what we tell them, but at least we will have made the best effort possible at capturing these effects.

Tom Steeger, as an EPA scientist, had been around long enough to know what to expect from parties with a vested interest in a particular product. "Posturing by the regulated community with EPA is a common practice and unfortunately science can be manipulated to serve certain agendas," he wrote to Hayes. Hayes, new to the game, viewed the actions of Syngenta and the scientific panel members as an affront and a betrayal. "Where is there [*sic*] accountability?" he asked Steeger in a July 31, 2002, e-mail. "They have already put out a press release saying that they cannot reproduce our work . . . When do they answer to this?" Replied Steeger: "Accountability is where it has always been, with those who share principles that distinguish right from wrong. Those principles do not dominate the corporate world."

If Hayes had any hope left that the EPA would carefully consider his findings in determining atrazine's fitness for the market, the Data Quality Act killed it. This obscure piece of legislation, written by an industry lobbyist and slipped into an appropriations bill just before President Bill Clinton left office, sounds on its surface like plain old common sense. It instructs federal agencies to issue guidelines that ensure "the quality, objectivity, utility and integrity of the information" the government disseminates. As implemented by the Bush White House in 2002, however, it gives outside groups the ability to legally challenge any information distributed by the federal government. And one of the first challenges made under the Data Quality Act was by the Kansas Corn Growers Association and the Triazine Network, farm trade groups that have received some funding from Syngenta. Filed in late 2002 by the Center for Regulatory Effectiveness (CRE), a lobbying outfit and think tank whose leader, Jim Tozzi, wrote the Data Quality Act itself, the petition went straight after an EPA reference to atrazine's endocrine-disrupting effects in frogs. "EPA's Environmental Risk Assessment accepts the endocrine effects allegedly shown by the Hayes Frog Tests as accurate and reliable," the petition states. Arguing that the EPA, under data quality guidelines, should not have "reached these conclusions and disseminated this information because there are no validated tests for detecting or measuring endocrine disruption in frogs," the petition asked that the EPA risk assessment "be corrected to state that there is no reliable evidence that atrazine causes endocrine effects in the environment."

By invoking the Data Quality Act, the makers and users of atrazine were insisting that regulators reject published peer-reviewed studies because the government lacked a standard test protocol for the research methodology that was used. The implications of the petition—which swept far beyond atrazine—troubled some scientists and environmentalists because the government lacks standard protocols to assess many types of studies, including epidemiologic research, prescription drug models, and molecular studies. "If accepted," wrote Jennifer Sass and Jon Devine of the Natural Resources Defense Council in a letter published in the January 2004 issue of *Environmental Health Perspectives*, "the CRE's arguments could jeopardize the government's ability to consider most published scientific research." Though Sass and Devine might have overstated to make a point, my review of Data Quality Act petitions filed with thirty-one federal agencies shows that big business has frequently used the legislation to "correct" the public record and squash regulation. Between 2002 and 2006, for example, the EPA received thirty-seven Data Quality Act petitions—twenty-three of which were filed by businesses or groups representing them. Among those, twenty had been responded to by the agency at the time of this writing, resulting in nine decisions favorable to industry. Chief among them was the agency's response to the CRE's atrazine petition, which, as described by *The Washington Post*, "effectively rendered moot hundreds of pages of scientific evidence" suggesting that atrazine disrupts hormones. Because Hayes's research methods were novel and a Syngenta-funded study failed to replicate his findings, the CRE was able to argue successfully that Hayes's data were not "reliable" under the new federal data quality guidelines.

Accordingly, the EPA biologist Tom Steeger's May 2003 report on the significance of the amphibian findings casts doubt on the significance of Hayes's findings, specifically citing that other labs were unable to reproduce the results. "Overall, the weight-of-evidence based on currently available studies does not show that atrazine produces consistent, reproducible effects across the range of exposure concentrations and amphibian species tested," Steeger wrote. "The current body of knowledge has deficiencies and uncertainties that limit its usefulness in interpreting potential atrazine effects." Hayes could not contain his disgust that the EPA would give any credence to the flawed Syngenta-funded studies submitted by his former panel colleagues. "I have little to no faith in the process as I understand it," Hayes wrote in an e-mail to Steeger on June 4, 2003. Hayes's analysis wasn't perfect, but the Syngenta-sponsored studies were riddled with basic problems—contaminated controls, high mortality, and poor animal husbandry—rare from Ph.D.s.

Normally, such studies would not be considered by the EPA. Yet the EPA accepted the results because the agency lacked established protocols for frog studies. In the e-mail to Steeger, Hayes told the EPA scientist he believed his former colleagues deliberately sullied their experiments. "I think that they started confirming our findings . . . and then intentionally diluted the data with garbage." A Syngenta toxicologist, however, insisted there was no such tinkering. "It does not help Syngenta to have these kinds of issues hanging out there," said Tim Pastoor, the company's head of global risk management. "We want to know what the answers are. We want to be able to responsibly steward the product. Without being able to resolve differences, it paralyzes the regulatory process and paralyzes our ability to make a clear and definitive statement on the health and safety of our product."

Hayes, however, is convinced this is exactly what Syngenta wanted—and what it got. On October 31, 2003, the EPA said it would allow continued sales of atrazine in the United States and announced that Syngenta had agreed to provide more amphibian data designed, specifically, to clear up the "uncertainties" raised by previous studies. Based on data submitted by Syngenta, the EPA concluded in late 2007 that "atrazine does not adversely effect amphibian gonadal development." According to the EPA,

"no additional testing is warranted to address this issue."
Hayes was outraged, but not surprised. "It's part of the
same thing: the EPA found a way to eliminate all the data
except Syngenta's," he told me. The result, Hayes said, is a
"deeply flawed" and limited review that does not adequately consider the scope of the environmental and
public-heath problems created by the use of atrazine. The
EPA limited its review to one specific question: Does
atrazine affect amphibian gonadal development? In doing
so, Hayes believes the EPA overlooked the more significant question: Is atrazine an endocrine disruptor that
poses potential and real harm to wildlife and human
health? Open, peer-reviewed literature, including studies
from Hayes's lab, surely suggest so. But "by asking limited
questions, of course the EPA gets limited answers," Hayes
said. "In my opinion, the whole process was set up to do
nothing."

Tyrone Hayes approaches his life with an uncommon zeal.
At times, he nearly seems to be in two places at once. In
order to present his research data at an important conference and attend his son's soccer game, he flew to Tokyo,
gave his talk, then immediately went back to the airport
and jumped on a plane for home. He does not like to slow
down. On a day I met with him at his Berkeley lab, he told
me his battered old pickup truck had run out of gas twice
in the preceding twenty-four hours. "I hate stopping for
gas," he explained with a shrug. "It's such a waste of
time." In the same vein, Hayes considers going to the bathroom an annoying interruption, which may help explain
why he invariably declines water when he dines. "It's not,"

he adds with a smile, "because I'm afraid of what's in it."

Though the supporters of atrazine have painted Hayes as an extremist with an antipesticide agenda, he is, by other measures, a highly regarded and skilled researcher. On a website (www.atrazinelovers.com) maintained by Hayes, he explains that his work is motivated by both a passion for science and a quest for social justice.

Citizens in lower socio-economic classes and, in particular, ethnic minorities are less likely to have access to this information, more likely to be employed and live in areas where they are exposed to pesticides, less likely to have access to appropriate health care, and more likely to die from what are already the number one cancers in men and women (prostate and breast cancer, respectively), with cancer now being the number one cause of death in the U.S.

Industry has increased efforts to discredit my work, but my laboratory continues to examine the impacts of atrazine and other pesticides on environmental and public health. My decision to stand up and face the industry giant was not a heroic one. My parents taught me, "Do not do the right thing because you seek reward . . . and do not do the wrong thing because you fear punishment. Do the right thing because it is the right thing." If I want to raise my children with the same philosophy, then I have to live my life in accordance with the way I direct theirs.

Where he diverges from other research scientists is in his outspokenness. He told me he feels sorry for those who can't—or won't—take a stand when it comes to something as sacred as scientific integrity. "Really, I don't know how they sleep at night," Hayes said. Yet he is the first to admit that his tenured position at Berkeley has given him a certain freedom to confront what seems to some to be science conducted in the private interest. "The whole atrazine reregistration process was more political than scientific," said Frank J. Dinan, a chemistry professor at Canisius College in Buffalo, New York. He was so flabbergasted that he decided to write up the EPA's findings for the National Center for Case Study Teaching in Science. Today, undergraduates all over the country can analyze the same data the EPA considered and debate what influenced the agency's decision. Dinan, though, has his own opinion. "It just was not a fair fight," he told me. "Syngenta hired all kinds of people to publish all kinds of things the EPA said was baloney and still they got atrazine reregistered."

Hired hands rode in on other fronts too. Atrazine's backers enlisted former senator Bob Dole to take their agenda to the Oval Office. Five weeks before the EPA announced it was reregistering atrazine, Dole lobbied a White House senior official and recommended that the EPA declare, without further delay, that atrazine was safe for continued use. A memo of Dole's talking points was obtained by the NRDC after it sued to force the government to release records related to the pesticide industry's atrazine-lobbying efforts. Dole took the position that scientific uncertainties "can always be addressed" after the reregistration is issued. Policy makers, it seems, heard his message. Despite outstanding lawsuits by environmental groups alleging the EPA failed to adequately consult the Fish and Wildlife Service on potential adverse impacts of the herbicide on endangered species, the agency followed Dole—and the industry's recommendation. It reregistered atrazine, while stipulating that research to clear up "scientific uncertainties" be continued. Other documents obtained by the NRDC suggest that the EPA for many months colluded with Syngenta and industry representatives to hammer out the scenario for atrazine's continued use. NRDC staff attorney Aaron Colangelo contends that the EPA and Syngenta negotiated atrazine risk levels out of public view and without regard for the state of the science. "What the EPA did with atrazine was like a new way of regulating," he told me. "They secretly worked it out with the manufacturer and negotiated a deal."

After the EPA officially sanctioned atrazine's safe use, Hayes took the gloves off, but not without warning at least one of his former panel colleagues, Texas Tech University associate professor Jim Carr, of his intentions. "I am very sorry for what has become of you," Hayes wrote to Carr on February 3, 2003, in an e-mail he shared with me, "but not sorry for what I have to do. I will be thorough, I will be ruthless. If I have a religion it is faith in science and acedemia [sic]. I believe that a few of us are given a gift. Trained to develop our intellect, to accumulate knowledge. With this gift, we have the responsibility to deliver truth. Our citizenship demands that we use this gift to teach and train others in our methodology and to pursue truth. Our stewardship demands that when given the opportunity, we use that knowledge to the benefit of the environment . . . I take my responsibility and my stewardship seriously. So serious that it exceeds sympathy for a former colleague who has forgotten these responsibilities. Forgotten where he came from, lost site [sic] of where he is going. I will deliver the truth. It is my responsibility as a scientist, as a teacher and as a citizen."

In November 2003, Hayes rocked the house at the normally genteel annual conference of the Society of Environmental Toxicology and Chemistry when he publicly accused Syngenta of subverting science to manipulate the EPA. He followed that up in 2004 and 2005 with fifteen trips to Minnesota where a group of legislators was building a case for why the state should be the first to do what the EPA didn't—restrict atrazine. Minnesota farmers use about two million pounds a year of the substance. But like most people in the United States, Minnesotans may not have a clear perspective on the implications. "I do a lot of work in Africa," Hayes told a state senate panel. "If I were to go to Uganda and tell these men that this water coming off their crop, which goes into this container that is their cooking, bathing, and drinking water, that this water chemically castrates frogs, I think they would immediately see that connection. In this country, we don't see the connection. We don't collect our water and take it back to the house in containers. It comes out of a faucet." But, Hayes cautioned, the government says it's okay for it to be coming out of a faucet at a level that is thirty times higher than what it takes to chemically castrate fish and frogs.

In the end, the Minnesota atrazine bills went nowhere, though to make sure of it Syngenta spent a bundle on St. Paul's fanciest lobbying firm and the EPA sent in its deputy director of pesticides to "correct the record" with respect to the soundness of its regulatory process regarding atrazine. Hayes wasn't fazed. In the spring of 2005, he shared the stage with Syngenta officials and industry-sponsored scientists at an atrazine conference he had spent two years organizing.

"I was happy that Syngenta showed up and at least appeared to be on the up-and-up," said Hayes, who looked as content as a kid catching frogs during the two-day conference in Iowa City. "I'm hoping it marks a new phase in our relationship. I think they realize they can't pull shit anymore because they know this guy is not going to let them get away with passing off half-assed science as good science."

He has continued his research on atrazine, and his most recent findings are no less disturbing than his first reports of sexually scrambled amphibians. In a January 2006 paper published online by *Environmental Health Perspectives*, the journal of the National Institute of Environmental Health Sciences, Hayes presented more evidence that atrazine both chemically castrates and feminizes male frogs. His lab produced identical hermaphroditic malformations in African clawed frogs by administering estrogen or blocking androgen during critical windows of development. As in his previous work, Hayes suggested that the induction of the aromatase enzyme caused by atrazine is a likely cause of the hormonal chaos behind the malformations. And he reminded his readers that frogs are not the only species affected in this way. Aromatase induction by atrazine, he writes, is "a mechanism established in fish, amphibians, reptiles and mammals (rodents and humans)."

More significant because of its real-world implications, Hayes's new research, also published in the same journal, shows that pesticide mixtures, the toxic brew fermenting in farmers' fields where frogs grow up, have stronger effects on amphibians than any pesticide alone. His data involved nine compounds—including atrazine—found at levels ranging from 0.1 part per billion to 10 or more parts per billion in water around a Nebraska cornfield early in the growing season. Hayes discovered that native northern leopard frogs raised in water with only one of the nine substances at 0.1 ppb appeared normal. But frogs exposed to all nine substances at 0.1 ppb each had a variety of physiological problems, including slow metamorphosis, retarded growth, and an increased susceptibility to meningitis caused by normally benign bacteria. Hayes wrote that these substances were not solely to blame for amphibian population declines. Instead, pesticide effects combined with nonnative predators, ultraviolet light, and global warming—all bad news for frogs—are teaming up to create a threat. "If you're a tadpole and your pond is drying up and your world is going away, that's a stressor," Hayes told me. "Stress hormones affect everything from reproduction to immune function. If pesticides are having their own impact on those same hormones, then you really have problems."

Hayes is continuing to study the effects of various combinations of pesticides. He believes regulators may be profoundly underestimating the risks of pesticide exposures by setting standards one chemical at a time. The U.S. Geological Survey shares similar concerns. "The common occurrence of pesticide mixtures, particularly in streams, means that the total combined toxicity of pesticides in water, sediment, and fish may be greater than that of any single pesticide compound that is present," noted Robert Gilliom of the USGS. Gilliom was the senior author of a sweeping 2006 report describing the occurrence of pesticides in the nation's waterways. Pesticides were seldom found alone. Rather, they almost always were detected in complex mixtures. Atrazine in particular was found together with deethylatrazine, one of its several degradates, in about 75 percent of stream samples and about 40 percent of groundwater samples collected in agricultural areas across the country. Of the fifty-one major river basins and aquifer systems surveyed by the USGS, most stream samples and about half of the well samples contained two or more pesticides—and frequently more. Concluded Gilliom, "Our results indicate studies of mixtures should be a high priority."

In August 2007, Representative Keith Ellison, a Minnesota Democrat, introduced legislation that would ban atrazine from national sale and distribution. Ultimately, however, the fate of atrazine use in the United States may rest with the courts. As of this writing, six class-action lawsuits against the makers and a distributor of atrazine are pending in an Illinois state court alleging atrazine is harmful at any detectable level, not just levels that exceed the EPA's safe drinking water standard of 3 parts per billion. The litigation, which seeks a variety of financial penalties, clearly has Syngenta's attention. The company posted a detailed response on a website called atrazinefacts.com. Under the headline "Atrazine Litigation Facts," Syngenta explains that the lawsuits, in part, rely on "scientific research which has not passed the litmus test of sound science: the ability to repeat tests with scientific and statistical confidence." The reference, of course, pertains to Hayes's research. But the site fails to mention how Syngenta and its proxies contributed to the uncertainty surrounding atrazine's potential effects, and how that uncertainty helps keep atrazine on the market. As will be seen, this same strategy has worked to keep questionable ingredients even closer to home in products ranging from cosmetics to computers.

WHAT PRICE BEAUTY? PHTHALATES AND YOU

On May 17, 1933, young Hazel Fay Musser's mother went to Byrd's Beauty Shoppe in Dayton, Ohio, to get spruced up for a PTA banquet at which she would be honored for her volunteer work. The blue-eyed brunette got a shampoo and a haircut. Considering the importance of the occasion, Mrs. Musser took a suggestion to touch up her eyelashes and eyebrows with dye—a popular cosmetic technique in an era when the application of mascara was both difficult and messy.

It turned out to be a devastating decision. Mrs. Musser reacted adversely to paraphenylenediamine, or PPD, a toxic dye used in a widely available touch-up product called Lash Lure. At first the Lash Lure treatment simply caused her eyes to sting, but the pain grew worse that evening and Mrs. Musser left the banquet early. By morning her eyes were swollen shut. Ulcers formed on her corneas, and for days her eyes seeped pus. When it was all over, Mrs. Musser had completely lost her vision. Her tragic case was among four published in the November 11, 1933, issue of the *Journal of American Medicine* documenting injuries from the use of Lash Lure. Others were reported in medical journals—before and after her experience. But it was a letter from her ten-year-old daughter, Hazel Fay, to President Franklin Delano Roosevelt that turned Mrs. Musser into a cause célèbre for the first cosmetics regulations in the country's history, the Food, Drug, and Cosmetic Act of 1938.

Throughout history, humans have used makeup to enhance what we have and fake what we don't. Egyptians brushed minerals on their faces to add color and definition. Greeks applied chalk and lead powders for a lighter look. Romans used foundation creams made of fat, starch, and tin oxide formulations that were as sophisticated as anything sold today. Colonial Americans concocted their own cosmetics based on formulas popular with Englishwomen in the 1600s and 1700s. In keeping with Puritan sensibilities, cosmetics were medicinal tools—not beauty preparations—used to clear the complexion, produce good color, and soothe sunburns. Women handed down their favorite cosmetics recipes to their daughters and passed them around to their friends throughout the nineteenth century. Except for "wicked" types spied in dance halls, saloons, and bordellos, most women of the era did not wear visible makeup.

Around the turn of the twentieth century, however, there began a decided shift in the morality of using makeup. Respectable women, many of whom previously dabbled in cosmetics on the sly, took to openly painting and powdering their faces. Then, as today, media images influenced consumer behaviors. In the early twentieth century, cosmetics became a leading source of advertising revenue in dozens of new magazines aimed at women eager to emulate the glamorous stars of stage and screen. The nascent beauty products industry challenged the Victorian view of makeup as a mask of deceit, promoting it instead as a mark of a woman's individuality in the modern world. As a marketing concept, it still holds today. In fact, the notion of physical transformation is what distinguishes cosmetics from drugs under federal guidelines.

In awkward regulatory language, the Food, Drug, and Cosmetic Act describes cosmetics as "articles intended to be rubbed, poured, sprinkled, or sprayed on, introduced into or otherwise applied to the human body . . . for cleansing, beautifying, promoting attractiveness, or altering the appearance." Included in this definition are a variety of products found in virtually every U.S. home: toothpaste, deodorant, shampoo, hair color, moisturizer, perfume, lipstick, nail polish, and eye and face makeup. The average adult applies or uses these products between twenty and twenty-five times a day.

While today's products rarely maim or kill—as was the case as late as 1938—the effects of using cosmetics that contain certain ingredients suspected or known to cause cancer, reproductive issues, and developmental problems is a concern the $60-billion-a-year U.S. cosmetics industry has been reluctant to face. In 2004, the European Union prohibited cosmetics makers from using any ingredients known to cause or strongly suspected of causing cancer and developmental or reproductive problems. But in the United States, where laws are less stringent, industry representatives prefer to frame the notion of product safety around short-term, acute consequences—think rashes—rather than hormone-disrupting or gender-bending outcomes that might be associated with long-term exposures to certain ingredients. Skin sensitivity is the number one product complaint documented by the Food and Drug Administration. "Of course, a rash to the person who has it is not minor, but in terms of public-health problems . . . cosmetics just don't appear on the radar screen," said Dr. Gerald McEwen, a lawyer and physiologist who retired recently as vice president of science for the Cosmetic, Toiletry, and Fragrance Association, or CTFA. In November 2007, the organization changed its name to the Personal Care Products Council, explaining that it was a better way to describe the business of its six hundred member companies.

McEwen, tall, gray haired, and nattily turned out as befitting a representative of the beauty-products industry, speaks with the authority of a man who long served as the trade group's liaison to the Cosmetic Ingredient Review, a panel of seven physicians and scientists who evaluate the safety of cosmetics ingredients. The cosmetics trade and lobby organization funds the review panel, which the FDA relies on to assess the safety of ingredients. In the through-the-looking-glass world of regulatory politics, the FDA uses the Cosmetic Ingredient Review to assess the safety of cosmetics ingredients, while the industry group claims that consumers can rest easy about cosmetics safety because the FDA closely monitors it. "The FDA has wide-ranging legal power to regulate cosmetics," states the council's website.

Unfortunately, that's a misleading statement at best. As the FDA's own website notes: "FDA's legal authority over cosmetics is different from other products regulated by the agency, such as drugs, biologics, and medical devices. Cosmetic products and ingredients are not subject to FDA pre-market approval authority, with the exception of color additives." In addition, the FDA is not authorized to substantiate product safety and performance claims. And it cannot require beauty-products manufacturers to register their operations or products, though some do it voluntarily. So while the FDA's mandate with respect to cosmetics and personal-care products is to protect the public health by ensuring that cosmetics are safe and properly labeled, it does not have the statutory power and resources to complete this public-health mission.

Through a request I made using the Freedom of Information Act, I learned that only thirty employees work in the FDA's Office of Cosmetics and Colors. The office's annual budget of $3.4 million is the same as it was nearly two decades ago and does not include funding for safety assessments. To put into context the low priority of this watchdog agency, consider this: the city office that controls the 989 traffic signals in Portland, Oregon, where I live, has forty employees and a budget of $22 million.

Before a chemist named Dr. John E. Bailey replaced McEwen as the top scientist for the industry group, he ran the Office of Cosmetics and Colors. His perspective on the industry while he was with the FDA is worth noting. "The cosmetic industry is sensitive to the image of an uncontrolled market where anything goes," Bailey told an interviewer in 1992. "They counter this image with well-established self-regulation programs. Part of the incentive for such industry policy is to avoid increased regulatory authority." So far, the strategy has worked wonders, allowing the industry to sell 11 billion personal-care products each year with essentially no oversight and no requirement to test for safety.

The Environmental Working Group, an advocacy organization funded by a number of progressive philanthropies and donors, analyzed the ingredients of 23,000 personal care products in 2007. The results were shocking: nearly 400 products contained chemicals prohibited from the same type of use by other developed countries, some 1,300 products contained ingredients for which the industry's Cosmetic Ingredient Review expert panel said there is insufficient data to determine whether they are safe in cosmetics, and more than 22,500 products contained one or more ingredients that had not been assessed for safety by the FDA or the industry's review panel.

About 10,500 ingredients are used in the product category of cosmetics, spanning everything from common table salt to chemicals linked to cancer, birth defects, and reproductive problems, according to the EWG analysis. Yet the FDA has banned or restricted only nine ingredients in the seven decades it has monitored cosmetics safety. By contrast, the European Union explicitly prohibits the use of more than 1,000 ingredients in beauty-care products, even though the majority on the list had never been used by the industry. Phthalates, a family of plasticizers linked to reproductive and developmental toxicity, have been showing up in biomonitoring studies at levels that perturb some scientists, and the public has shown increasing concern. The twenty-seven-nation EU banned dibutyl phthalate (DBP) and diethylhexyl phthalate (DEHP) from personal-care products in 2004. The move forced nail polish manufacturers to remove DBP, which makes the polish soft and flexible, if they wanted to continue doing business in Europe. Some manufacturers—including Estée Lauder (Clinique and M.A.C.), Procter & Gamble (Max Factor and Cover Girl), L'Oréal (Lancôme Paris and Maybelline), and Avon—also removed DBP from their U.S. nail polish lines. But they described the move as a business decision and not a safety issue. "We didn't have an option if we were to continue to market in Europe," said Tim Long, a Procter & Gamble spokesman, about the company's decision to remove DBP. "In the U.S., we product-tested the reformulation and found that consumers liked it better." Other manufacturers, including Revlon, said their nail polish never contained DBP.

Despite the EU's restrictions and a move in 2005 by the state of California to list DBP and DEHP as chemicals that require label warnings when present in a product at certain amounts, the FDA took no action. "I cannot say definitely that there are no human health effects," said Dr. Linda M. Katz, director of the FDA Office of Cosmetics and Colors. But the data the FDA has seen so far "does not establish an association between phthalates in cosmetics posing a significant health risk." The FDA's position mirrors the conclusion on phthalates reached by the Cosmetic Ingredient Review, which has found three phthalates, including DBP, DEHP, and the widely used diethyl phthalate (DEP) safe for use at current product concentrations.

The FDA's inaction, coupled with stiffer regulations in the European Union, inspired a coalition of health, consumer, and labor groups to form a national Campaign for Safe Cosmetics. The campaign began by asking cosmetics and personal-care products companies to sign a pledge to remove toxic chemicals from all products they sell. "We're trying to change the nature of the dialogue and the debate to say no amount of a suspected or known carcinogen or reproductive toxicant should be considered okay in a cosmetics product," said Janet Nudelman, program director for the Breast Cancer Fund in San Francisco, a founding organization of the campaign. While hundreds of smaller companies and so-called natural products niche marketers have accepted the invitation and made the pledge (including the Body Shop, Avalon Natural Products, Kiss My Face, and Zia Natural Skincare), campaign organizers soon learned that asking nicely didn't get them very far with the biggest names in the cosmetics game. "The conventional companies have told us: 'We're already safe. We're not doing anything wrong. We're not exposing customers,' " said Nudelman. "They just flat-out refuse to acknowledge that there's any kind of learning curve as we gain more scientific knowledge."

So the campaign, which as of this writing included thirty thousand individual members, stopped asking and started acting. Startling advertisements appeared in national newspapers. One featured a little girl—her lips smeared with bright red lipstick—under the headline: "Putting on makeup shouldn't be like playing with matches." Thousands of letters were written and phone calls made to cosmetics companies and retailers imploring them to sign the campaign's safe-cosmetics pledge. OPI Products, a leading company known for offbeat nail polish names such as Melon of Troy and All Lacquered Up, was subjected to an advertising spoof in *Variety* and public protests in dozens of cities depicting OPI's Miss Treatment USA, a beauty queen in an evening gown whose nail colors included I Didn't Know Carcinogens Came in Coral and Phthalates of Pearl.

Why target OPI? Months before the actions, campaign coordinators traveled to Los Angeles to meet with company executives. Nudelman told me the meeting was cordial, but the OPI executives refused to remove toluene, formaldehyde, and DBP from its products. All three substances are on California's Proposition 65 list of chemicals known to cause cancer and reproductive toxicity, and the EU has banned DBP, found in some nail polishes at a concentration of up to 15 percent. Indeed, OPI was making DBP-free nail polish for EU consumers while continuing to put DBP in polish sold in the United States. Campaign organizers decided they could build a strong ad around such information. Moreover, it seemed the time had come to challenge a market leader. The strategy paid off. In late 2006, OPI announced it would begin removing DBP from its products. Two other nail polish companies, Orly International and Del Laboratories, which makes the popular Sally Hansen drugstore brand, immediately followed suit. Finally, as the pressure from the Campaign for Safe Cosmetics continued, OPI also agreed to remove toluene and reduce the use of formaldehyde-forming substances in its products. (Rather than argue about how much of a toxic chemical is okay to use, the campaign would prefer, of course, that companies completely eliminate questionable chemicals from their products.)

Reformulating is expensive, but not impossible, as companies such as Avalon Natural Products can attest. Avalon markets the Alba, Avalon Organics, and Un-Petroleum brands of personal-care products. Morris Shriftman, senior vice president of marketing, said the company went through a painful but ultimately rewarding process in order to comply with the Campaign for Safe Cosmetics' toxics-free pledge. "It was a mission we called consciousness in cosmetics," said Shriftman. The company, under new ownership at the time it was approached by the campaign, struggled at first. "What do you do when you learn about something that can be dangerous, problematic, and expensive? We had a rude awakening."

But once enlightened, the company got going. It reviewed its products and formulated a grid of ingredients standards, Shriftman said. Avalon did not use any of the substances already prohibited in the European Union. Then it looked at ingredients for which safety data might be lacking. Once it was satisfied in that regard, the next step was to make sure the company was avoiding the use of non-renewable petroleum-based ingredients. And then, despite a lack of consensus about its hazards, parabens, widely used preservatives in items such as shampoos, moisturizers, shaving gels, and toothpaste, got the boot from Avalon's product lines. "We learned enough to accept that parabens are endocrine disruptors," said Shriftman. "And we decided there was no reason to have endocrine disruptors in our products."

In the end, he said, the company invested $1 million in

product reformulations but has gotten it back many times over. "Our retailers are getting the message and the consumers are getting the message," said Shriftman. "More importantly, we are delighted the company has a higher purpose. We're in the skin care and beauty business, and you could say that's a bullshit, vanity business. And in many ways, you could say it's true. But we found a deeper value in what we do. And the world of consciousness in cosmetics has been very good from a business point of view." The company's sales climbed to $40 million. Then, in December 2006, Shriftman and his partners proved that consciousness in cosmetics can provide financial as well as ethical returns when the company announced it was being acquired by Hain Celestial Group, a natural and organic food and personal-care conglomerate, for $120 million.

Among the most outraged over the cosmetics industry's just-trust-us attitude are some of its best customers. I met a few of them on a balmy spring afternoon in the rarefied surroundings of Northern California's Marin Civic Center, an aging Frank Lloyd Wright building perched on a knoll not far from San Pablo Bay. Five teenage women occupied the council chamber seats reserved for the Marin County Board of Supervisors. Each delivered a statement—with the blessings of the supervisors—warning cosmetics manufacturers that they had better come clean about hazardous ingredients and sketchy labeling. The high school students were, by all appearances, an unlikely group to be indicting the makers of makeup and personal-care products. Eyeliner and mascara framed their bright eyes. Lipstick outlined their plump mouths. Blush highlighted their youthful glow. Foundation camouflaged their adolescent blemishes. Each had begun buying cosmetics while still in middle school and used makeup to feel more confident and pretty. Not one had a feminist quarrel with this artifice. Nor did any want to forgo face paints or body lotions. In the bloom of their youth—and their buying power— they represented a slice of the industry's most sought-after consumers. But after learning about a lack of safety requirements that gives the cosmetic industry leeway to put harmful ingredients into beauty products, these girls were angry. "We are drenching our skin in toxins that accumulate in our bodies," said Victoria Ruff, a seventeen-year-old who looks remarkably like a young Angelina Jolie. "Manufacturers know about the dangers, but most have yet to do anything. This is intolerable," and, according to Victoria, should not stand. Which is why she and her fellow teens, each a member of a safe cosmetics club at her high school, were imploring cosmetics manufacturers to clean up their acts. The teens had drafted a cosmetics bill of rights demanding accurate and truthful product advertising and labeling, the exclusive use of toxics-free ingredients, better enforcement of federal safety guidelines, and premarket safety testing of product ingredients. "We can stand up to corporate America and say, 'We're unhappy with the ingredients in cosmetics products,'" said Sasha Hoffman, a winner of Miss Teen USA World who has modeled professionally. She used the issue of safe cosmetics as her platform during a pageant competition and dumped her favorite national-brand makeup, M.A.C., in favor of Dr. Hauschka, a narrowly distributed line that does not contain toxic ingredients. "As teenagers, we do have a voice," she said. "We are the industry's best customers."

It's fitting that California, the state where the glamorous entertainment industry is based and whose governor is no stranger to makeup and movie sets, would be the first to enact a Safe Cosmetics Act. The legislation, which took effect in 2007, signals a turning point in the push for more information and better accountability. Pressure from consumer and health groups is one thing, but the unprecedented legislation, signed by Republican governor Arnold Schwarzenegger, means manufacturers must disclose to the Department of Health Services product ingredients listed on state or federal registries of chemicals that cause cancer or birth defects. The law gives California authorities the right to investigate the health impacts of chemicals in cosmetics and requires manufacturers to supply health-related information about the ingredients. Emboldened by new regulations in the European Union and the victory in California, the safe cosmetics movement is now waging a David vs. Goliath battle that is moving to the federal level, where a powerful cosmetics lobby has been accustomed to getting its way.

"My judgment with respect to cosmetics is that the cosmetics lobby has effectively gutted any semblance of oversight," said Oregon senator Ron Wyden. As a young Democratic congressman with a progressive bent, Wyden conducted Capitol Hill hearings in 1988—the fiftieth anniversary of the Food, Drug, and Cosmetic Act— spotlighting loopholes in cosmetics regulations. "People all over America get up in the morning and from the time they kick off the sheets until the time they climb back into bed—fourteen, sixteen, eighteen hours later—they're slathering these products on themselves," the senator told me. "The assumption of citizens is that somebody in government is making sure these products are safe, and the reality is that nobody's home."

After the 1988 congressional hearings, Wyden set out to strengthen cosmetics regulations. He wanted Congress to require that cosmetics ingredients be tested before they hit the market and to give the FDA access to safety and consumer complaints lodged with the industry. Wyden also proposed making FDA registration of cosmetics products and ingredients mandatory because few manufacturers and distributors did so voluntarily. But his draft legislation—similar in scope and language to reforms contained in an unsuccessful cosmetics safety bill introduced by Missouri senator Thomas F. Eagleton, a Democrat, in 1973— got heavy industry resistance and went nowhere. "We weren't asking cosmetics makers to tell the government, heaven forbid, their trade secrets," said Wyden, his sarcasm making his frustration evident nearly twenty years later. "We just wanted to know a little bit about what goes into these products so government could move quickly if it needed to."

Unfortunately, not much has changed since Wyden's big push. After the Campaign for Safe Cosmetics reported finding lead concentrations of .02 to .65 parts per million in 61 percent of thirty-three name-brand red lipsticks it tested, three U.S. senators—John Kerry, Barbara Boxer, and Dianne Feinstein—asked the FDA to test the same lipsticks and a variety of others. "If the FDA reaches the same conclusion as the Campaign for Safe Cosmetics, we respectfully encourage the FDA to take immediate action to reduce consumers' lead exposure to [*sic*] lipstick and other cosmetic products," the senators wrote in a letter on November 19, 2007. A few weeks later, the FDA announced it had "decided to allocate the resources" for independent testing of a selection of lipsticks on the market. At the same time, however, the agency said it was not valid for the Campaign for Safe Cosmetics to compare the lead levels in lipstick (for which there is no safe limit set by the FDA) with the FDA-recommended level for lead in candy. The FDA's reason wasn't particularly reassuring for any woman who happens to lick her lips or eat something while wearing lipstick. According to the agency, lead in lipstick is less of a worry than lead-tainted candy because lipstick is intended for topical use and "is ingested in much smaller quantities than candy."

Along with its name change in 2007, the Personal Care Products Council launched an interactive consumer website, acknowledging a growing desire on the part of consumers to know more about the products they use. At the time of its debut, the website, www.cosmeticsinfo.org, listed more than fifteen hundred ingredients. "Consumers have an increasing desire to understand the ingredients in cosmetics and personal-care products and are more proactive in that kind of information gathering," said Lisa Powers, vice president of information. However, critics said the trade organization left out vast amounts of damning findings. "It's disturbing that rather than responding to legitimate concerns about toxic ingredients, the industry's main trade group changes its name and launches a PR campaign," said Stacy Malkan, the author of the 2007 book *Not Just a Pretty Face: The Ugly Side of the Beauty Industry*. Malkan, a cofounder of the Campaign for Safe Cosmetics, told me that what can be found on the trade group's website pales in comparison to the information consumers can access on a database called Skin Deep, which sorts the ingredients in nearly twenty-five thousand products against toxicity information. That website, www.cosmeticsdatabase.com, is maintained by the Environmental Working Group.

Given our exquisitely intimate relationship with the products that make us clean and pretty, the lack of oversight regarding what's in them is especially discomfiting. Every day, spanning a lifetime, we put a variety of personal-care products all over our skin—the body's largest organ, stretching twenty-one feet in the average person. As the toxicologist Marc Lappé pointed out, the skin is a schizophrenic, denying access to the body of some things and actively encouraging the passage of others. Until the latter half of the last century, the skin was seen as an almost impermeable barrier. Today, however, it is well established that the same construction of lipids and proteins that makes the skin resistant to water also makes it vulnerable to the entry of certain agents. With this knowledge, the pharmaceutical industry has gained FDA approval for dozens of medicinal patches in the last twenty-five years. There are patches to prevent motion sickness and smoking, patches to prevent pregnancy and provide hormone replacement, and patches to treat overactive bladders and high blood pressure. Many patches contain penetration enhancers. So do many cosmetics, but the toxicity of certain ingredients in cosmetics and personal-care products isn't subject to the same scrutiny as pharmaceuticals.

Among cosmetic ingredients, phthalates are particularly controversial. The widely used plasticizers are found in everything from blood bags to intravenous tubing, toys to garden hoses, plastic bags to food packaging, pharmaceutical products to pesticides, detergents to solvents, and lubricating oils to beauty products. From a molecular standpoint, low-weight phthalates are used in the stuff we rub and spray on ourselves. They fix the scents in fragrances and emulsify ingredients. High-weight phthalates make plastics soft and pliable. Phthalates, produced in quantities of over six billion pounds a year, are everywhere. We breathe them, absorb them through our skin, and ingest them in food and water. Our bodies break down phthalates in a matter of hours—unlike some persistent pollutants that hang around in body fat for years. Yet the CDC has shown that phthalates exposure is widespread in the U.S. population and detected at levels ranging over three orders of magnitude, which means we're constantly exposed to a daily soup of the chemicals. In 2005, the CDC found elevated levels of phthalates in children, a finding also noted in an earlier German study. Using phthalates restrictions enacted by the European Union and California as a model, Senator Dianne Feinstein introduced a 2007 bill to ban toys and child-care products containing more than 0.1 percent of any of six types of phthalates. The need for such federal action, which passed as part of the Consumer Product Safety Improvement Act of 2008, was underscored when the Washington Toxics Coalition released a report showing that playthings the group purchased from Toys"R"Us, Wal-Mart, and Target contained phthalates at levels ranging from 28 percent to 47.5 percent. "Europe and California have already stepped forward and made sure that toys laden with phthalates are kept away from the hands and mouths of young children," said Feinstein. "America's parents should be able to have the same peace of mind that the toys they buy for their children are safe." To that end, Washington governor Christine Gregoire signed legislation on April 1, 2008, to enact the most stringent toy safety standards in the nation—including strict limits on phthalates.

At least five different phthalates are present in cosmetics and personal-care products, according to the FDA, and scores of products that do not list phthalates as a separate ingredient contain them anyway. That's because fragrance formulations, which are heavily dependent on phthalates, are listed simply as "fragrance" on ingredient labels. And any ingredient claimed as a trade secret is also exempt from labeling.

The lack of labeling information for parents is especially frustrating for them after researchers led by a team from the University of Washington found higher levels of phthalates in the urine of infants and toddlers whose caregivers applied baby lotions, powders, and shampoos to the children's skin and hair. The research, published in February 2008 in the journal *Pediatrics*, suggests that baby-care products contain a variety of phthalates that are absorbed through the skin. It also found that infants under the age of eight months, who are the most vulnerable to potential adverse effects because of their still-developing endocrine and reproductive systems, are the most highly exposed. Noting that there is no U.S. requirement that baby-care products be labeled as to their phthalates content, the authors of the study urged caution. "If parents want to decrease exposures, then we recommend limiting the amount of infant care products used and not to apply lotions or powders unless medically necessary."

Toxicologists know from rat studies that the reproductive tract of the developing male fetus is the most sensitive to phthalates exposures, though exposure to DBP and some other phthalates can disrupt male reproduction at all stages of life if given at a high-enough dose. Dozens of results have shown that high doses of phthalates affect the endocrine system in utero, thereby inhibiting male hormones and causing a range of abnormalities from birth defects to infertility to cancer. Scientists named the phenomenon "phthalates syndrome." As research advanced, primarily led by the National Institute of Environmental Health Sciences, it became increasingly clear that lower exposures to some phthalates, including DBP, had toxic effects on the male reproductive tract of rats. At the same time, rat studies demonstrated that exposures to more than one phthalate acted additively, increasing the negative changes seen in the lab animal. "Really, the critical question for humans is: Do we ever get doses that could induce the effects we see in animals?" said Paul Foster, a senior research fellow in toxicology at NIEHS who conducted several phthalates studies. "When you start to see effects in animals at lower levels and, at the same time, higher exposures in humans, your comfort factor is not as strong." His colleague, the environmental epidemiologist Jane Hoppin, is less sanguine, calling the extent of human exposure troubling. "No one expected such high levels," she said.

Scientists postulate that the prenatal exposures that produce phthalates syndrome in rats might provide some clues about the reasons for increases in human male reproductive abnormalities. Rates of testicular cancer rose about 50 percent in the last quarter of the twentieth century, while sperm quality measurably declined. At the same time, baby boys were increasingly born with hypospadias, a condition in which the opening of the penis is found along its shaft instead of the tip, and cryptorchidism, a failure of the testes to descend fully. In fact, hypospadias and cryptorchidism are the two most common birth defects in American boys.

One recent theory holds that all these disorders originate in the womb and comprise a condition called testicular dysgenesis syndrome, a cause of which could be in utero phthalates exposure. But testing the theory—or any that involves the effects of chemicals on humans—is one of the most complex tasks facing science. Toxicologists use animals as surrogates because they cannot ethically administer doses of chemicals to a human (let alone one who is pregnant) and wait to see what happens. So epidemiologists try to tease out the real-world effects of exposures that have already occurred. Invariably, toxicological and epidemiological results are fraught with uncertainty, which gives interests on either side of any particular debate generous wiggle room. In the case of phthalates, for example, personal-care manufacturers claim that animal studies showing reproductive toxicity are at exposure levels that aren't relevant to humans, and that human studies are too few to bear weight. An Italian study found that DEHP can inflame the uterus of expectant mothers and may contribute to premature birth. A study of men from a Massachusetts General Hospital fertility clinic observed a correlation between exposure to DBP at environmental levels and reduced sperm quality. The results were consistent with earlier research findings by this same team. And the University of Rochester found a statistically significant correlation between phthalates exposure and abdominal obesity and insulin resistance in adult U.S. males.

On the one hand, all of these studies were preliminary in nature. On the other hand, animal studies raise significant concerns, and the lack of conclusive human data should not be mistaken for evidence that a substance is safe. Debates like this can go on for years—and often do. But in the spring of 2005, Dr. Shanna Swan, an epidemiologist from the University of Rochester School of Medicine and Dentistry, weighed in with a finding on phthalates that, though preliminary, carried tremendous public-health implications. Swan, a tiny woman whose cutting-edge research on hormone-disrupting compounds tends to create enormous controversy, found that women with higher levels of exposure to four different phthalates— including three sometimes found in cosmetics—were more likely to have baby boys with the worrisome characteristic known as shortened anogenital distance (AGD). Published in the peer-reviewed journal *Environmental Health Perspectives*, the study of 134 babies demonstrated that the more a mother was exposed to phthalates, the greater the chance her son would be born with a shorter perineum, the tissue between the genitals and anus also known as AGD.

Shortened anogenital distance is a sensitive indicator of the prenatal phthalates exposure that causes gender-bending effects in male rats. Swan wanted to determine if its presence in a baby boy correlated with a mother who had higher phthalates exposure during pregnancy. As it turns out, there was a strong connection. While none of the boys had clearly deformed genitals, in the 25 percent of mothers with the highest levels of phthalates exposure, the odds were ten times higher that their sons would have a shorter than expected anogenital distance. Most startling of all, Swan saw adverse effects at lower doses than those seen in rat studies, suggesting greater human sensitivity to phthalates than previously believed. Indeed, a quarter of U.S. women have phthalates exposure levels associated with the shortened AGD observed by Swan. A follow-up study by researchers from the University of California and the EPA estimated that the phthalate levels found by Swan would have been caused by exposures approximately one hundred times lower than EPA's current safety thresholds.

Swan must follow the boys with shortened AGD to see if they develop problems over time. And she cautions that the preliminary study needs to be expanded and repeated. Nevertheless, it's a significant step. Scientists have now seen in humans the first inklings of the same devastating effects on the male reproductive tract observed in rats— and at even lower levels of exposure. With that, it becomes even more critical to determine the routes of human exposure to phthalates in order to stop them. I asked Swan if she surmised that a woman's use of cosmetics was connected to the chance of dysfunction or disease in the reproductive system of her baby boy. Her answer was cautious but sobering: "I can't say for certain what the source of the exposures are. But I can say the data support the hypothesis that prenatal phthalate exposure at environmental levels can adversely affect male reproductive development in humans."

Olivia James had never heard of hypospadias until 1997, when her obstetrician showed her the opening on the penis of her newborn, Darren. The opening was underneath the shaft of Darren's penis, about midway between the tip and the scrotum. In the most severe cases of hypospadias, the opening is found almost at the base of the penis. Regardless of severity, hypospadias often requires that men sit to urinate. In some cases, hypospadias makes sexual intercourse impossible. At nine months, Darren had surgery to correct the condition. Years later, when James learned that cosmetics contained chemicals linked in animal studies to the occurrence of hypospadias, she began to wonder if her regular use of makeup and beauty products—including foundation, powder, concealer, eye shadow, mascara, hair relaxers, lotions, and nail polish—could have contributed to Darren's condition. "I'll never know for sure," said James, a financial analyst for Dow Jones in Princeton, New Jersey, and a former professional model. "But I feel guilty thinking the cosmetics I use might have hurt my son." James believes regulators do not take cosmetics safety seriously and she is angry that it's impossible for consumers to know for sure what ingredients beauty products contain. "They just don't think makeup is all that important," she said. Unless the United States matches the standards set in the European Union, advocates of safer cosmetics say Europeans are going to be far more confident than U.S. consumers about the safety of the cosmetics. "We think U.S. women's health is worth erring for on the side of safety," said the Breast Cancer Fund's Janet Nudelman.

In certain ways, the mounting pressure for safer cosmetics recalls the atmosphere leading up to the passage of the Food, Drug, and Cosmetic Act of 1938. Then, like today, the market was robust. Cosmetics were one product that proved Depression-proof. Sales soared throughout the 1930s, leading C. C. Concannon, then chief of the Commerce Department's Chemistry Division, to conclude that the cosmetics boon was due to "a psychological factor— the conviction that the race is to those who feel fit, and a large part of feeling fit is looking fit. So far as cosmetics act as an aid to that end of really making people feel better, they have ceased to be a non-essential and have become a necessity."

To everyone's great relief, cosmetics consumers no longer face worries about the kind of mutilation endured by previous generations. Today, however, there are legitimate concerns about how cosmetics ingredients affect the body's most intricate systems. Phthalates research clearly demonstrates the public-health implications, even though many questions remain. And strangely enough, the genitals of a few baby boys may provide some of the very first answers.

 UP IN FLAMES: POLYBROMINATED DIPHENYL ETHERS

In the 1970s, the makers of bromine chemicals piggy-backed on the success of plastics and polyurethane materials by introducing a cheap and effective way of increasing their fire resistance. The appearance of polybrominated diphenyl ethers couldn't have come at a better time for all concerned. The bromine business was losing a major product—a leaded gas additive—because of the switch to unleaded fuels based on environmental and health concerns. Meanwhile, consumer-goods manufacturers had all but abandoned traditional materials such as wood, metal, and wool in favor of petroleum-derived plastics and polyurethane foam. But they were struggling with how to keep their new, highly flammable products from burning quickly. Staples of the modern home and office, such as carpets, drapes, and upholstered furniture, were prone

to catch fire, and the way they burned left less time for es-
caping from and fighting flames.

PBDEs seemed like the perfect answer. To begin with,
they have applications for everything from televisions,
mattresses, upholstered furniture, and insulation to computers, car trim, carpet pads, and drapes. Not only are they
inexpensive, but they are easily blended into a product
during manufacturing—imagine a baker stirring chocolate
chips into cookie dough. And they work like a winning
science-fair entry: high heat or flames trigger PBDEs to release bromines that rob the air of the oxygen that fires
need to stay alive. Ideally, PBDEs, which are one branch
of a large and chemically diverse family of bromine flame
retardants, stop combustion. Even if something starts to
burn, the fire-inhibiting properties of PBDEs delay the
time it takes for flames to flash over to other combustible
materials. According to the Bromine Science and Environmental Forum (BSEF), an industry group run by the international public relations firm Burson-Marsteller, the use of
all types of brominated flame retardants saved an estimated 280 lives in the United States in 2000. And it's without question that these widely used chemicals have greatly
increased the fire safety of the products in which they are
used, preventing deaths, injuries, and property losses. But
there's a downside to the use of PBDEs that has some of
the world's foremost environmental scientists and toxicologists on edge. PBDEs escape from materials in which
they are mixed. The result: they contaminate our air and
environment, and because we breathe and eat them, they
are building up in humans at astonishing rates.

Wipe your finger along the baseboard of your living

room or the miniblinds in your office, and the dust you pick up assuredly contains tiny amounts of the flame retardants. Roll up the lint from your dryer's filter, and the downy wad in your hand contains the stuff, too. Take a bite of that steak on your plate or a gulp of milk from your cup and you're ingesting traces of PBDEs along with your protein. In the three decades since manufacturers began adding PBDEs to plastics and foam, these chemicals have contaminated people, animals, and places all over the world.

Levels of PBDEs in human fat, blood, and breast milk in North America—the largest user of these flame retardants—are ten to one hundred times higher than those reported for Europe and Asia. And the levels are continuing to increase. While most research has revealed total PBDE levels between 4 and 400 parts per billion in human blood and milk, a 2005 study of fatty tissue donated by a New York plastic surgeon discovered PBDE concentrations of nearly 10,000 parts per billion in a thirty-two-year-old man, and 4,060 parts per billion in a twenty-three-year-old woman. The two were among fifty-two liposuction patients whose tissue was collected and analyzed for PBDEs. The group had the highest median and mean exposure levels of PBDEs ever recorded. "We have now seen [PBDE] concentrations in some people higher than where we have seen effects in experimental animals," said Dr. Linda Birnbaum, director of the EPA's Experimental Toxicology Division. "It's amazing and a little bit scary."

This rapid buildup of PBDEs has stunned scientists and spurred new research into how humans are exposed. Clues come from the very persistent industrial solvents known as polychlorinated biphenyls, or PCBs. They are chemical cousins of the flame retardants. As with PCBs, there are 209 different molecules, also called congeners, in the PBDE family. Each molecule, or congener, is assigned a number; the bigger the number, the larger the molecule. PBDEs closely resemble PCBs in their chemical structure, and both types of chemicals are produced commercially as mixtures. Widely used as insulating fluids and coolants in electrical equipment and machinery from 1929 until they were banned in 1977, PCBs last for years in the environment and get into people through their diet. So do PBDEs. But diet alone does not explain the high levels measured in humans, and scientists suspect that contaminated house dust contributes to exposures. In 2004, a study funded by the National Institute of Standards and Technology and the Environmental Protection Agency analyzed floor dust and clothes dryer lint collected from seventeen homes in Washington, D.C., and Charleston, South Carolina. It found PBDEs in every sample at concentrations ranging from 700 to 30,100 nanograms per gram. This raised concerns that toddlers, who spend a lot of time playing or sitting on the floor, are most at risk to PBDE exposures from dust.

In 2005, researchers from the University of Toronto and Environment Canada added strength to the theory that house dust serves as a significant exposure pathway—and that toddlers are disproportionately exposed. According to their computer modeling, infants receive the largest exposure through breast milk. But toddlers, children, teens, and adults get their biggest doses of PBDEs through inadvertent ingestion of house dust. Even more remarkable, the modeling shows that toddlers with a high dust intake rate can have exposure levels almost a hundred times higher than average if they live in a home in which PBDE levels are elevated.

A 2006 case study of a California family's actual PBDE exposures bears out some of the patterns predicted by computer modeling. The study, published in *Environmental Health Perspectives*, measured concentrations of PBDEs in two sets of blood serum samples collected three months apart. The research was commissioned by *The Oakland Tribune*, which published a series of stories about body burden in March 2005. A Berkeley family of four—Michele Hammond, Jeremiah Holland, and their children, ages five years and eighteen months—volunteered for the study. Notably, the Hammond Holland family did not use many products known to contain PBDEs. They eschewed common household cleaners and pesticides, had no wall-to-wall carpeting, and did not own any large new appliances. Nevertheless, Hammond and Holland each had levels of PBDEs approaching U.S. median concentrations for adults. And their children had exposures that rivaled the maximum amount found in U.S. adults. Rowan Hammond Holland, the youngest child, was exclusively breast-fed for his first six months and was still breast-feeding during the study period. His total PBDE exposure of 651 parts per billion was, the authors noted, "uncomfortably close to body burdens associated with adverse effects on reproduction and neurodevelopment in laboratory animals."

No one can say for certain how PBDE exposures are affecting humans, young or old. But research published in

2007 by a team under the direction of the EPA's Birnbaum makes the case that house cats may be sentinels for humans, especially the littlest ones. "Think about a cat's behavior," said Birnbaum. "They're on the floor; they're on the furniture. They get dust on themselves. They ingest it through grooming. Well, little kids are all over the floor and furniture, and they're frequently mouthing things, including their hands."

Thirty years ago, about the time that PBDEs were introduced, veterinarians began noticing an uptick in feline hyperthyroidism, the most common endocrine disorder in cats. It causes rapid weight loss and leads to secondary problems with the heart and digestive system, greatly diminishing an animal's quality of life. Notably, the disease is associated with cats who live indoors. Birnbaum and her research team wondered: Could the epidemic of thyroid disease in pet cats be associated with PBDEs in house dust?

Their pilot study showed that PBDE exposures in older cats with hyperthyroidism were three times higher than the levels in younger cats without it. Some cats in the study had PBDE exposures one hundred times higher than those measured in humans, and even the healthy cats had levels twenty times higher than humans. Researchers theorized that the difference in human and feline exposure levels may be owing to the fact that children outgrow the behaviors that lead to high dust intake, but cats never do. Birnbaum told me that more studies are needed to pin down a correlation between PBDEs and feline hyperthyroidism, but that it makes sense to study chronic PBDE exposure in cats because they share the same environment with humans. Moreover, as the study notes, cats and humans are the only mammals that have a high incidence of hyperthyroidism.

As with hundreds of other widely used chemicals, human toxicity data on PBDEs does not exist. According to the EPA, however, studies suggest potential concerns about the liver, the thyroid, and human development. Not so surprisingly, these toxic effects are similar to those associated with PCBs.

Developing fetuses and young children, who are at risk for the greatest exposures, also are at risk for the greatest harm: studies on lab animals show that fetal exposures to PBDEs can affect brain development, causing problems with learning and movement after birth. If PBDE levels continue to double every two and a half years, within ten years the average person may have levels of PBDEs similar to those shown to cause developmental damage in mice. Highly exposed people already have PBDE levels within this threshold. No scientific studies have confirmed that humans are being harmed even now, but it cannot be ruled out. The EPA's Birnbaum told me she suspects that developmental PBDE exposures are lowering thyroid hormone levels—crucial for normal brain development—and putting significantly more children, perhaps millions of them, at risk for permanent learning and developmental deficits. This type of shift is called a population effect and it has significant societal costs.

Low-level lead exposures had similar implications, Birnbaum explained. These exposures robbed each individual of a few IQ points, which is considered a marginal effect. But applied to a large population, the exposures shifted the IQ distribution curve downward. With thyroid hormone levels, Birnbaum continued, "There is a distribution in the population and some people are naturally at the high end or the low end of the normal range." And if PBDEs are causing thyroid hormone levels to drop? That may not be a problem for the person who has more than the average within the normal distribution of the population, Birnbaum said. But for the person who is in the low-normal range to begin with, a further decrease means the risk of permanent deficits.

Whispers about problems with PBDEs began in the late 1990s when Swedish researchers discovered, quite by accident, that exposure levels were increasing exponentially in breast milk samples. While looking for evidence of a different contaminant in archived samples collected over twenty-five years, researchers found a jaw-dropping sixty-fold increase in the concentrations of PBDEs in breast milk. Between 1972 and 1997, the concentrations doubled every five years. Experts did not think that even the highest levels detected (approximately 4 parts per billion) would cause harm to a nursing baby, nor did they discourage breast-feeding because of the findings. Nevertheless, officials in Sweden, which has one of the highest rates of breast-feeding in the world, treated the startling information about the increase in PBDE exposures with great urgency. The government turned to its official Chemicals Inspectorate, which reviewed the chemicals and submitted a proposal in early 1999 to prohibit the use of PBDEs. At the same time, manufacturers such as IKEA and Volvo moved to voluntarily phase out their use of the most commonly detected PBDEs, replacing them with other flame retardants. Immediately after Sweden discontinued using PBDEs, total levels in breast milk began falling and dropped 30 percent between 1997 and 2000.

Spurred by the findings in Sweden, researchers across the globe soon were looking for PBDEs in people, and finding them everywhere they turned their gaze. Canadian experts discovered PBDE levels in the breast milk of Vancouver women that were fifteen times higher in 2002 than they had been ten years earlier, doubling every 2.6 years as compared to five years in Sweden. Not only were the levels rising more quickly in Canadian women, but the average PBDE levels were at least ten times higher than in Sweden. In the United States, it was even worse: a 2003 study of nursing mothers in Texas revealed levels of flame retardants in their breast milk that were ten to one hundred times higher than levels seen in Europe. That same year, a study from California showed that PBDE exposure in the blood and fat tissue of Bay Area women had more than tripled since the introduction of commercial PBDEs three decades earlier. And the exposure news hasn't gotten any better. As of this writing, scientists know that PBDEs have been detected in all North Americans ever tested for them at levels between 5 and 10,000 parts per billion. Taken as an average, North Americans have PBDE exposure levels approximately ten times higher than people in Europe or Asia. The reason, experts believe, is that more PBDEs have been used in North America than in any other place in the world.

When word of PBDEs in Swedish breast milk first broke, some news reports focused on all the scary-sounding pesticides, chemicals, and heavy metals expressed in the milk. And while it's true that breast milk commonly contains these chemical pollutants, the reports missed the salient point: researchers, in general, analyze breast milk not because they believe it's a threat to a baby's healthy development; rather, they monitor breast milk because it's a handy and relatively inexpensive way to study chemical exposures in a population. Pollutants that persist in the environment, such as DDT, PCBs, dioxin, lead, and mercury, hitch themselves to the fat molecules in breast milk, just as they do to a body's other fatty tissues. But it's less invasive to collect breast milk from lactating women than it is to excise a tissue sample or draw the large quantities of blood needed for biomonitoring studies. Authorities ranging from the American Academy of Pediatrics to the Centers for Disease Control to the World Health Organization continue to encourage breast-feeding because it is still the absolute best thing for mother and child. Indeed, the nutritional and health benefits of breast milk have been shown to help counter the damaging effects of a baby's in utero exposure to toxic substances. (Remember, persistent pollutants cross the placenta, so babies receive their first exposures to toxic chemicals while they're developing in the womb.)

Laura Mittelstadt, thirty-nine, an active, outdoorsy part-time bookkeeper from Beaverton, Oregon, nursed each of her two children until their first birthdays, knowing that her breast milk contained traces of PBDEs. Just after her oldest, Nathan, was born in 2003, Mittelstadt participated in a breast milk study of forty Pacific Northwest women. It pinpointed her total PBDE exposure at 54 parts per billion, which was slightly above the study median. It's a cruel and creepy truth that studies confirm breast-feeding always lowers a woman's levels of toxic contaminants because she is, in effect, off-loading some of her toxic body burden directly to her nursing infant. Studies also show that firstborn children receive higher exposures through breast milk than their younger siblings.

Despite her exposure levels, Mittelstadt is confident the benefits of breast-feeding outweigh the known contamination and attendant risks. "We live in a world where there are chemicals you are exposed to all the time," she said. "If I didn't have a single chemical residue in me, that would be incredible." The news of her PBDE exposure levels did not, for a second, dissuade her from breast-feeding Nathan or her daughter Katie, born in 2005. Both children, she happily reports, are healthy, and Mittelstadt knows that nursing them boosted their immune systems and provided nutrients far superior to formula feeding. Breast-feeding her children, she told me, was a simple, easy thing she could do to give Nathan and Katie a good start in life.

If a breast-feeding woman is healthy, the benefits to her baby extend from sustenance to survival. Nursing babies are better able to fend off nasty viruses and infections, and fewer of them experience sudden infant death than formula-fed infants. At the same time, breast milk contains prophylactics for later-in-life illnesses such as asthma, allergies, diabetes, and cancer. And it even seems to make kids smarter. Moms, families, and society benefit from breast-feeding, too. Nursing mothers decrease their risks for breast and ovarian cancer and osteoporosis. A family pockets thousands of dollars by not having to buy infant formula. And, to complete the circle of sustainability, society saves on public-health costs and the environmental impacts of making formula and disposing of the empty containers.

As we've seen, the problem is not with breast milk. Instead, it's the overall body burden of chemicals in humans, of which toxics-tainted breast milk is but one marker. For nursing mothers, however, it can be a source of personal outrage—and a rallying cry. "There should be nothing more basic than a mother's right to provide clean and healthy breast milk for her child," said Mary Brune, a founder of the grassroots group Making Our Milk Safe (or MOMS), in a 2006 interview with the online magazine *Grist*. "Whether you're a blue-state mom or a red-state mom, nobody wants their kids exposed to toxic chemicals. This is a human issue—something we all, as parents, are confronted with." Brune and three other nursing mothers founded the activist group in 2005 to help raise awareness of the need to eliminate toxic threats. Breast milk contamination makes a strong case for why we should be finding ways to reduce exposures to toxic chemicals, said Kim Hooper, a senior scientist who runs a laboratory for the Department of Toxic Substances Control at the California Environmental Protection Agency. "When breast milk speaks," he said, "people listen."

By late 2003, the bromine industry had heard enough about the problems with PBDEs. The flame retardants were regarded by some scientists as "the PCBs of the twenty-first century," and no industry wants to have its product compared to a chemical so harmful it was banned by the Toxic Substances Control Act itself. The two chemical families share similar chemical structures and fates. Both have low solubility in water; are persistent in the atmosphere, water, and soil; exhibit strong adsorption to sediment and sludge; are transported to and contaminate remote locations (including the Arctic); and bioaccumulate. Scientists predict that the concentration of PBDEs in animals will soon surpass levels of PCBs, which are thankfully on the decline.

In terms of toxicology, "PBDEs act like PCBs in animal studies," said Dr. Arnold Schecter, a professor of environmental and occupational health studies at the University of Texas in Dallas and one of the leading PBDE researchers in the United States. Both cause problems with neurological development in animal experiments, and PCBs cause neurological damage in humans. "Based on what we know, I would consider PBDEs to probably be about as toxic as PCBs," he told me.

Like PCBs, PBDEs have been very widely used. When the EPA issued final regulations banning the manufacture of PCBs in 1979, the agency estimated that 150 million pounds were dispersed throughout the environment and an additional 290 million pounds were in landfills. Brominated flame retardants have been used just as liberally. Worldwide demand in just one year, 2003, was nearly 447 million pounds, according to a report by Maine state agencies.

With the use of PBDEs becoming increasingly difficult to defend, Great Lakes Chemical, the only U.S. manufacturer of PBDEs, announced it would cease making two of the three commercial mixtures of the flame retardants at the end of 2004. The company, now known as Chemtura following a 2005 merger with Crompton, discontinued the two substances, known as Penta and Octa, after the European Union and California instituted bans, and other states were lining up to do the same. Penta, considered the more toxic mixture, had been used mainly in mattresses, seat cushions, and other upholstered furniture. Up to 95 percent of it was applied in North America. Octa was found primarily in housing for fax machines and computers, car trim, telephone handsets, and small kitchen appliances.

The voluntary halt to production of Penta and Octa was clearly a step in the right direction, but California's Kim Hooper describes it as "a gimme." Together, Penta and Octa accounted for about a quarter of the North American market for PBDEs, and Chemtura had a viable replacement product for Penta ready to go. The real fight, said Hooper, is being waged over the most widely used PBDE formulation, known as Deca, which is almost entirely composed of congeners that contain nine to ten bromine atoms. At first, Deca was not suspected of building up in human tissues. The larger molecules in the mixture were thought to be too stable to be a toxic threat. Once researchers understood how rapidly PBDE exposures were rising, they began to wonder if Deca was somehow contributing to the problem. Now, a growing stack of scientific studies suggests Deca can break down, or degrade, into more toxic and bioaccumulative forms of PBDEs like the ones no longer in production.

The industry refuses to disclose demand for the commercial Deca mixture. BSEF North American program director John Kyte told me the information is considered proprietary. However, a 2007 report by the Maine Department of Environmental Protection and the Maine Center for Disease Control and Prevention states worldwide demand for Deca was nearly 113 million pounds in 2003. About 80 percent of all Deca is used in housings for television sets and computers. Based on the rate of U.S. TV sales in 2003 and the amount of Deca used in an average TV, one report estimated that a total of 37.8 million pounds of Deca are contained in TV housings sold each year. The next largest use of Deca is in textiles, including commercial-grade furniture upholstery for offices, cars, planes, and trains. It is also applied to some mattresses, but it is not generally used in apparel or bed linens. Of course, stricter flammability standards—perpetually pushed by manufacturers—always help sales, which may help explain the bromine industry's strong resistance to all efforts that would restrict Deca. The good news is that environmentalists managed in 2008 to halt new fire safety standards under consideration by the International Electrotechnical Commission (IEC) for external, open-flame ignition of electronics with low internal operating temperatures (think ink-jet printers, telephones, and fax machines). While the standards would have opened fresh revenue streams for millions of pounds of brominated flame-retardant products, including Deca, opponents successfully argued that new requirements would unnecessarily open homes, offices, and schools to chemical exposures. Nevertheless, the bromine industry, while acknowledging that it makes many of the alternatives to Deca, pitches its old standby with the fervor—and script—of a telemarketer.

"Deca is well-suited to the applications for which it is most commonly used—plastic housings of TVs and other electrical equipment, electrical connectors, wire and cable

covering, automobile and airplane components, and textile backcoating—providing excellent ignition resistance
and flame retardant properties," Kyte told me in an e-mail.
"In addition, Deca does not affect the material properties
of the plastics in which it is used, and it is highly efficient,
requiring a relatively low volume of material to provide
high levels of ignition resistance. Alternative flame retardants can require higher volumes of the selected flame
retardant to achieve similar levels of ignition resistance."

Of this rallying for Deca, environmental and public-
health advocates take a more cynical view. The industry,
said Michael Belliveau, executive director of the Environmental Health Strategy Center in Bangor, Maine, sacri-
ficed two marginal products in order to save a favorite. "It
is frighteningly clear what the manufacturer has done is
strategically withdraw Penta and Octa in hopes of defending the market for its largest commodity product, Deca."

The same scientists who talk about the similarities between PCBs and PBDEs also like to point out one big difference. Levels of PBDEs are increasing in people and the
environment while levels of PCBs are going down. Even
with Penta and Octa out of production, experts foresee
total PBDE levels continuing to build up in people, animals, and places before they start to decline. For one
thing, the use of Penta and Octa in durable goods has created large reserves. Penta and Octa are in products that
people keep around for a long, long time. (When was the
last time you replaced your sofa, your electric mixer, or
your car?) In addition, they were added in copious quantities (Penta can account for as much as 30 percent of a

product by weight and Octa as much as 18 percent). And, because of the way they were blended in during the manufacturing process, they leach out during a product's entire life cycle—from manufacturing, through the years spent in a home, office, or car, and finally, when the poor old thing is put to rest in a landfill or incinerated, potentially forming toxic dioxins and furans.

And then there's the issue of Deca. Of particular concern is the potential for Deca degradation via debromination—a process by which bromine atoms are sequentially removed from an organic compound, resulting in a smaller, lower-brominated molecule. These lower-brominated congeners have the potential to be more persistent and more bioaccumulative than their larger parent chemical. Testing indicates that Deca debrominates in sunlight. It also degrades via debromination in bacteria found in sewage sludge, which is sometimes used as soil amendment, and in soil where there is no oxygen, such as under water. Evidence suggests that the fate of Deca would be the same in the environment as it is under laboratory settings used for testing, though the rate at which these reactions occur would be slower. In short, the more scientists learn about Deca degradation, the grimmer the picture gets. When I first interviewed the EPA's Linda Birnbaum about flame retardants in 2003, she told me Deca was not a substance that gave her the same kinds of worries as Penta and Octa. By 2006, her assessment was markedly different. "Our concerns about Deca have been increasing over the past couple years," said Birnbaum. "More and more data shows it breaks down in the environment. The questions are, how much of it is breaking down and how toxic are those breakdown products?"

As of this writing, the makers of Deca have no intention of taking it off the market. "We have recently completed a review of all the studies to date indicating the potential for Deca to degrade," John Kyte told me in an e-mail. "While it is correct that Deca can be made to degrade in laboratory settings, generally under extreme conditions, this does not appear to have any direct correlation to the real environment or to the types of congeners being found in the environment." In other words, the industry does not accept that the debromination of Deca, under real-world conditions, is a significant risk to human health or the environment. And it's been willing to go to great lengths to dissuade the EPA from making that determination, too.

In the summer of 2007, the American Chemistry Council used its considerable influence over the Bush administration in order to silence an award-winning scientist who has been outspokenly in favor of eliminating Deca. The toxicologist Deborah Rice, an EPA retiree who had moved on to the Maine Department of Health and Human Services, was removed as chair of an expert scientific panel reviewing the safety of Deca after the ACC, the industry lobbying group, complained that Rice was "a fervent advocate of banning Deca." Earlier in 2007, Rice had testified before her state legislature in favor of a Deca ban and also was quoted in media reports as saying there is enough scientific evidence to warrant banning the substance. EPA officials noted they removed Rice from the panel and excluded her comments from the expert-panel review because of "the perception of a potential conflict of interest."

Environmental and public-health advocates, incensed by the EPA's actions, accused the agency of perpetuating a dangerous double standard under the Bush administration, which has allowed many pro-industry experts to serve on agency scientific panels. "When the government removes top scientists from positions because they express concerns over potential health risks from industrial chemicals—at the same time leaving dozens of scientists with direct ties to the chemical industry on review panels—something is very wrong," said Jane Houlihan, vice president of research for the Environmental Working Group. To make the case, EWG analyzed seven EPA external review panels and found seventeen panel members with direct or potential conflicts of interest, including employees of companies who made the chemicals under review, and scientists whose work was funded by industries with a financial stake in the panel's outcome. Take, for example, the chemical industry consultant Betty Anderson, who chaired a panel that has recommended the EPA weaken by a factor of three the safety standard for dibutyl phthalate (DBP), a chemical discussed in the previous chapter. Anderson had no particular expertise in the toxicity of DBP to qualify her as panel chair. However, Anderson's employer, Exponent, was simultaneously under contract with the ACC's Phthalate Esters Panel to discredit Shanna Swan's key epidemiological study, which found a correlation between everyday exposures and a shortened anogenital distance in baby boys.

In another example of jaw-dropping audacity, Dr. Heather Stapleton, an assistant professor of environmental chemistry at Duke University, heard a BSEF scientist presenting at a 2006 worldwide conference argue that Deca is safe under real-world conditions. Stapleton, a young scholar conducting cutting-edge research in the fate and biotransformation of persistent organic pollutants including PBDEs, could hardly believe her ears. Not only was the presentation intellectually dishonest in claiming that the scientific literature did not support the theory of environmental Deca degradation, the industry group attempted to support its claims using excerpts from a recent paper Stapleton had written summarizing the significance and extent of Deca debromination. The trouble is, they grossly misrepresented her report. Stapleton said she "was absolutely livid. They completely ignored all the references that Deca was susceptible to breakdown and all the instances where it would be environmentally relevant and made it seem like Deca was just fine," she told me. It was as if, she said, they had not even read her paper. Stapleton's report pointed out a litany of concerns about Deca debromination. Among them: the dust found in houses, offices, and cars is full of the stuff, and when it is exposed to sunlight the potential for debromination is high; some fish appear capable of debrominating Deca when they metabolize it, and studies suggest humans might do the same; and concentrations of Deca measured at 10,000 ppb in house dust mean that toddlers are likely to ingest very high levels of the flame retardant. Added Stapleton, "It's unfortunate BSEF takes the position it does, but that's what it's paid to do."

The public relations people at Burson-Marsteller who run BSEF are well schooled in how to defend and promote questionable products. They honed their skills working for tobacco companies and still use tactics employed in that fight. BSEF, for example, funds studies that support the continued use of Deca. It attacks science that does not fit the industry's agenda. And it downplays the dangers of exposures to PBDEs at levels found in people.

The scientific researchers who made up the audience at the brominated fire retardants workshop aren't easily swayed by industry-funded spin. But the same cannot be said of the public and policy makers—the real targets of the message. In 2007, Washington State became the first in the country to pass legislation requiring Deca to be phased out of all consumer products. It took three tries: similar legislation had failed under intense lobbying by the bromine industry the two previous years. A long list of environmental and health organizations, including the Washington State Nurses Association and the Washington chapter of the American Academy of Pediatrics, always favored the phaseout. And by the second year two state agencies, the departments of ecology and health, supported the plan after conducting a two-year study that included a review of the scientific literature on Deca, a cost-benefit analysis of a phaseout, and an assessment of alternative substances. The report from the state agencies weighed in at 310 pages. Supporters said they had the votes to pass the bill in 2006, but the senate failed to bring it to a vote after industry lobbyists "grandstanded information that didn't tell the whole story," said Robert Duff, director of the state Office of Environmental Health Assessments.

Particularly frustrating, said Duff, was the notion floated by the bill's opponents that a Deca phaseout was overkill—an unreasonable action that pandered to environmental extremists and put people and property at greater risk from fires. "It was dishonest," said Duff. An earlier plan for an outright ban on Deca had not had state agencies' backing until the bill was rewritten to embrace what Duff called an "off-ramp approach." The revised bill called for a gradual phaseout of Deca and gave the state the opportunity to cancel it if safer flame-retardant alternatives could not be identified—hardly a course of action that would put people in jeopardy to appease a handful of radicals.

To the contrary, after the exhaustive review of the scientific data on Deca, it was clear, said Duff, that the state should take responsible action to reduce exposures immediately. "We don't know for sure how much Deca is breaking down and contributing to the accumulation of Penta and Octa congeners," he said. "But as long as we have viable alternatives, why should we wait decades for more data that tells us just how much Deca is to blame?" Widespread human exposures to Deca make restricting it a public-health priority, said Duff. Using a risk assessment equation, "risk equals exposure times toxicity," Duff continued. "So even if the toxicity seems pretty small, the exposure cannot be any larger. Deca is everywhere, in every facet of our lives. Every single person is exposed, and this causes the equation to change. The potential risk could be enormous, even with mild toxicity."

In straightforward language, the authors of the Washington PBDE chemical action plan acknowledged the uncertainties surrounding the flame retardants, writing that state agencies developed their recommendations "after a thorough consideration of what is known and what is not known. We believe these recommendations represent prudent public policy, and that the suggested actions are commensurate with the risk involved, both to human health and the environment as well as to Washington businesses. What we want to avoid is adopting a policy that allows the continued build-up of PBDEs in our bodies and in the environment as we try to resolve the unknowns."

This approach, however, was seen as anything but pru-

dent by the industry-funded BSEF. Its chairman and
North American director fired off a response to the director of the Washington Department of Ecology that lam-
basted the work of the state agencies.

> Fundamentally, the Draft Final Plan makes its recommen-
> dations based on what [the agencies] claim is NOT known
> about Deca-BDE, rather than on what is known.
>
> Given the fact that Deca-BDE is clearly the most stud-
> ied and analyzed flame retardant in history, brominated or
> not, it is hard to understand how or why [the agencies] ar-
> rived at their conclusions and recommendations regarding
> Deca-BDE. A lack of knowledge or availability of reliable,
> published scientific information on Deca-BDE might ex-
> plain the "precautionary principle" approach being taken
> by [the agencies] but, in this situation, there is no lack of
> information—more information exists about Deca-BDE
> than any other flame retardant.

Washington State's Duff agrees that Deca is well studied, "but I've always found this particular argument amus-
ing because the fair amount of data that's been generated
on Deca indicates a problem. It doesn't exonerate it."

In another example of truth-shading, the bromine industry frequently argues that the European Union conducted extensive risk assessments of Deca and concluded
it does not pose a risk to humans or the environment. In
actuality the EU's risk assessments stopped short of issuing Deca a clean bill of health, advising that additional
study was needed to monitor the possible formation of
more toxic and bioaccumulative products that may result
from Deca degradation. Furthermore, critics of the EU

risk assessments say they were inadequate because they failed to consider the growing evidence that house dust is probably the main source of Deca exposure for most people. Adding to the controversy, a European Commission science advisory committee urged risk reduction measures for Deca, writing that emissions to the environment may pose serious problems in the future.

Fundamentally, the argument over Deca in Washington State and every jurisdiction weighing new rules is not about the strengths or shortcomings of particular risk assessments. Instead, it turns on whether policy makers accept the premise that risk assessment methods, based on conventional epidemiology and toxicology that underestimate the environmental links to certain diseases, should continue to be the benchmark for determining what is best for human health and the environment. Comparing, say, the recommendations of the Washington plan to phase out Deca with a course of more study and no further risk reductions makes the differences between precautionary and risk-based approaches starkly apparent. While the former errs on the side of protecting human health and the environment when viable alternatives are available, the latter leans toward continued use of a potentially toxic substance in the absence of definitive proof of harm.

As we've seen, risk-based approaches—in contrast to a better-safe-than-sorry approach now embraced in the EU—yield standards that put business interests ahead of public health. Federal agencies such as the EPA are hamstrung by a toothless Toxic Substances Control Act that allows the use of tens of thousands of chemicals without knowledge of their toxicity. So state and local jurisdictions are acting on their own to restrict questionable substances. As of this writing, ten states have banned Penta and Octa flame retardants, including Washington and Maine. Legislation to phase out Deca is pending in the eight other states with Penta and Octa bans. Significantly, California Assembly Bill 706 would prohibit all forms of brominated and chlorinated flame retardants. The chemical industry spent an estimated $10 million in a successful bid to table the legislation in 2007, though it remains eligible for further action. "Chemicals of this sort should be regulated at the national level," Winston Hickox told *The Wall Street Journal* in October 2003, when he was California's EPA secretary. When they are not, Hickox said, "the state must protect its citizens' health."

Washington State's Rob Duff told me that states' actions on PBDEs reflect the thinking of one of the luminaries of epidemiology. Duff pointed me to a quotation by Sir Austin Bradford Hill, who is perhaps best known for designing the smoking and lung cancer trials that linked tobacco addiction to patients' later-in-life problems with cancers and coronary heart disease. Back in 1965, during an address to the Royal Society of Medicine, Hill set the stage by stating, "All scientific work is liable to be upset or modified by advancing knowledge. That does not confer upon us a freedom to ignore the knowledge we already have, or to postpone the action that it appears to demand at a given time."

Scientific discoveries that some contaminants alter gene behavior at extremely low doses and that high-dose experiments do not predict low-dose effects have rendered traditional risk assessments woefully inadequate tools for prescribing public policy. These discoveries, considered far out when first discussed a decade ago, are now part of the scientific mainstream. Yet such points are difficult to get across to lawmakers and the public—especially when the EPA is silent and the industry offers soothing reassurances that substances such as Deca are as threatening as blue sky on a summer's day.

"I truly believe that a lot of the people on the industry side—the scientists and the producers—don't have a concern for low-level exposures," Duff continued. "They take the attitude that these tiny levels are part of life. From a public-health point of view, that doesn't cut it. My job is to be preventative. We prioritize things when we see them rising in the environment and when we have concerns for our kids. That's why we need to act on Deca."

In an interview with the *Fort Worth Star-Telegram*, John Kyte, the BSEF spokesman, said the detection of flame retardants in people does not mean that any harm is being done. "I think that is overblown," said Kyte. "We're evolving toward this detection-equals-danger mentality, when in science it always has been, and it still is, the dose determines the danger. And if you don't have the dose at a certain level, you don't have the danger. But what we're doing is evolving to sort of a public perception that the presence of something alone equals a danger. And that's just not common sense, and it's not good science."

While the the bromine industry insists that efforts to place restrictions on Deca are unfairly based on what is not known—an argument that confuses scientific uncertainty with no evidence of harm—manufacturers and retailers are already abandoning Deca. Companies such as Dell, IKEA, Apple, Sony, and Hewlett-Packard have for years demonstrated their ability to meet fire safety standards without the use of Deca, and many others are following their lead. As of July 1, 2006, the European Union restricted the use of PBDEs and five other toxic substances in electronics under the RoHS Directive, which restricts the use of certain hazardous substances in electrical and electronic equipment. However, Deca, the industry's darling, was exempted by the European Commission in a move that caused the European Parliament to launch a rare legal challenge. Nevertheless, manufacturers who quit using Deca are glad of it: on June 29, 2006, the European Commission decided Deca wouldn't do under its new rules for electronics products because the Deca mixtures contain prohibited lower-brominated congeners. Then, on April 1, 2008, the European Court of Justice ruled in favor of the European Parliament's challenge to the Deca exemption, banning Deca in electronic products effective July 1, 2008. As it stands, many businesses see no upside to using Deca, even without such broad restrictions in the United States.

A few clicks around the websites of leading computer manufacturers reveal how these companies are embracing environmental stewardship as a cornerstone of corporate responsibility. The environment page at Dell tells how its use of all brominated flame retardants has been eliminated in line with the precautionary principle and international treaties on priority chemicals. Not to be outgreened, Hewlett-Packard says it eliminated PBDEs a decade ago and plans to remove all brominated flame retardants from external case parts in new products. Apple too considers PBDEs to be twentieth century, boasting that it scrubbed the use of brominated flame retardant years ago and explaining how many of its products are "made of inherently flame retardant aluminum and polycarbonate plastic."

The move away from Deca hasn't been quite as thorough in televisions. According to a 2005 investigation conducted for the Lowell Center for Sustainable Production at the University of Massachusetts Lowell, a range of substitute materials that would eliminate the need for Deca are available. To use them would add from $4.40 to $7.50 to the cost of a twenty-seven-inch TV set that sells for roughly $300, excluding one-time switching costs to manufacturers for things such as new molds. The paper notes that rules in Europe have pushed the market away from Deca. "If market drivers existed in the U.S.," the authors wrote, "nearly all manufacturers have the technology and know-how to meet the demand for Deca-free products that meet strict fire safety standards." Some already are using it. Deca is absent from the housings of TV sets made by the industry giants Panasonic and Sony. Other large manufacturers such as Samsung and Sharp do not add the Deca flame retardant, though some may be in recycled plastics used in the making of new products.

The pressure on companies to reduce and eliminate chemicals of concern comes from all quarters, said Richard Liroff, executive director of the Investor Environmental Health Network in Arlington, Virginia, and a former senior fellow in the toxics program at World Wildlife Fund. "Companies that do not understand toxic hazards in their products and who do not take steps to reduce or eliminate them face the risk of disruption to their supply chains, exclusion from markets, damage to their reputation, foregone profits, and toxic tort litigations," he wrote in a 2005 article for the *International Journal of Corporate Sustainability*. On the other hand, he continues, innovative and entrepreneurial companies that are "first movers" in producing safer products can gain competitive advantage, reduce operating costs, increase profits, and enhance shareholder value.

To Laura Mittelstadt, the Oregon mother who experienced the luxury and curse of finding out the extent of PBDE contamination in her breast milk, the toxic traces found in all of us are a problem that cannot be fixed in one state or one country. In the long run, Mittelstadt said, preventing more toxics from tainting the world we share has got to be a global effort. "I sent a letter to the Oregon legislature supporting a ban on PBDEs," she said. "But when you shop at Costco and see pears from Peru, or travel to China and see how all our stuff is coming from there, you start to realize that whatever we do has to be done around the world. It does no good to ban a chemical in one place if it's used in another."

6 # THE GOODS ON BAD PLASTIC: BISPHENOL A

D r. Fred vom Saal wears his passion on his sleeve, a rarity among those who put on lab coats for a living. A professor of developmental biology and a researcher at the University of Missouri in Columbia, vom Saal is an expert—some say the foremost expert—on the effects of bisphenol A, the starting material for a polycarbonate plastic used for reusable food and beverage containers and an epoxy resin that lines most metal food cans. Tall and thin, with puppy-dog eyes and neatly combed hair that falls in either direction from a pencil-straight side part, vom Saal, who is in his early sixties, reminds me of television's Mr. Rogers—sans the sweater. But in looks alone do the similarities begin and end. For unlike Mr. Rogers, vom Saal can be pointedly harsh—at least when it comes to bisphenol A and the industry that makes more than six billion pounds of it a year. Those who espouse the

American Chemistry Council's official position on bisphenol A—that it poses no risk to human health—"are blatantly lying or complete idiots," said vom Saal. Scores of animal studies conducted in the past ten years demonstrate that bisphenol A, a compound that mimics the female hormone estrogen, causes an array of consequences at minuscule doses. And in humans? Because bisphenol A, or BPA, leaches from containers into our food and beverages, virtually everyone in the developed world is exposed to the chemical within the dose range that causes devastating problems in test animals. "There's no reason to think that what's happening in animals wouldn't occur in humans," said vom Saal. "When you look at molecular mechanisms at the cellular level, there is essentially no difference in the way that mouse and rat cells respond to bisphenol A and the way that human cells respond to it."

More than three dozen of the world's best brains on bisphenol A agree with him. In an unprecedented consensus statement remarkable for its simple, forceful declarations, thirty-eight scientists—all of them experts in researching the health effects of bisphenol A or similar endocrine-disrupting compounds—laid out their chilling conclusions. Nearly all of us (95 percent) have blood levels of bisphenol A within the range "that is predicted to be biologically active," based on animal studies conducted with low doses of the chemical. The problems seen in laboratory animals are "a great cause for concern with regard to the potential for similar adverse effects in humans." And these adverse effects—sure to cause shudders in any parent who has eaten canned food or given a child a polycarbonate baby bottle or sippy cup—correspond with recent trends in human diseases. The list includes increases in breast and prostate cancer, increases in urogenital abnormalities in male babies, a decline in semen quality in men, early onset of puberty in girls, metabolic disorders including Type 2 diabetes and obesity, and attention-deficit/hyperactivity disorder (ADHD).

Especially insidious is that outcomes are apparent only long after exposure to bisphenol A. "These developmental effects are irreversible and can occur due to low-dose exposure during brief sensitive periods in development," concluded the panel in the consensus statement published online by *Reproductive Toxicology* in August 2007. And although developing fetuses and young children are most vulnerable, the authors noted, "This does not diminish our concern for adult exposure, where many adverse outcomes are observed while exposure is occurring." Areas where problems from adult exposures may arise include the immune system and insulin metabolism.

Remarkably, however, you won't find any of these low-dose concerns by using the high-dose toxicology methods employed for assessing risks. This, combined with obfuscation of the science, political pressure, and ineffectual chemicals policies, means no regulatory agency had ever done a thing to lessen human exposure to bisphenol A—until Canada banned it from baby bottles just as this book was going to press.

Take a quick inventory of your surroundings and you'll begin to understand the significance of bisphenol A to the chemical industry and to consumer society. Without it, you cannot have polycarbonate plastic—the stuff of baby bottles, bicycle helmets, eyeglass lenses, water cooler jugs, and bullet-resistant barriers. Reusable food storage containers and beverage holders—the hard-plastic type popular for hiking and the gym—are made from it, too. Bisphenol A–derived epoxy resins line most food and beverage cans in your pantry and seal imperfections in your teeth. They coat bottle tops, water supply lines, electrical equipment, and adhesives. So useful are polycarbonate plastic and epoxy resins that bisphenol A production increased nearly fivefold between 1980 and 2000 among manufacturers such as Bayer Material-Science, Dow Chemical, General Electric, Hexion Specialty Chemicals, and Sunoco Chemicals.

It's not surprising the chemical industry would vigorously defend bisphenol A. Since 1997, when vom Saal published his first study, the industry has sponsored thirteen large-scale studies on low-dose effects. None found problems. By comparison, out of 163 low-dose studies funded by government and principally carried out at universities—most on a smaller scale—93 percent found evidence of harm. Dr. Rochelle Tyl, the director of toxicology for RTI International who conducted several of the industry-sponsored studies, told me that the different outcomes are not based on "who pays for a study, but the nature of the studies." University studies, "where you see all these strange and wonderful doses," typically focus on hazards, or the intrinsic capacity to do harm, said Tyl. In contrast, she said, studies paid for by industry are interested in determining the risks of exposures. "I get personally and professionally offended when someone suggests that who pays for the study determines the outcome."

And yet you need barely scratch the surface to see that the chemical industry has a fierce interest in protecting the status quo. To acknowledge that low-dose effects are real would mean calling into question current regulatory standards set by tests that are blind to the behavior of endocrine-disrupting chemicals. In the 1980s, when the EPA calculated that humans could safely be exposed daily to 50 parts per billion of bisphenol A per kilogram of body weight, the formula was based on a toxicological model that posits impacts will always worsen as exposures increase. The model also assumes the converse: that impacts will always diminish as exposures decrease. In other words, the dose makes the poison.

In the laboratory, here's how that works: traditional toxicology experiments test three to five doses of a substance spanning from low to high doses. Toxicologists start at the highest dose chosen and continue to lower doses until they find the point where experimental animals show no differences from control animals. This is called the No Observed Adverse Effect Level, or the NOAEL. Although the goal of a study design is to find a dose that causes no effect, toxicologists instead might find the Lowest Observed Adverse Effect Level, or the LOAEL. When these animal studies are used for risk assessments to determine a "safe" level of exposure in humans, additional safety factors are used. Toxicologists divide by up to 10 to account for conversion of LOAEL to NOAEL. They then divide by up to 10 again to account for uncertainties in extrapolating animal research to humans. Finally, they might divide by up to 10 a third time to account for sensitive individuals such as children or the elderly. Each factor builds in additional safety. The final number is called the reference dose, or acceptable daily intake. Based on traditional toxicology calculations, the reference dose for bisphenol A, 50 parts per billion per day, was assumed to be safe even though it was untested.

All this would be fine if bisphenol A produced a linear dose-response curve that started at the lowest point on the left side of a graph and rose diagonally across the page. But in study after study, it does not. Instead, bisphenol A, like other endocrine-disrupting compounds, produces a strange-appearing dose-response curve. These can be shaped like a U or an inverted U. In the case of bisphenol A, it is the latter. And what it shows is that medium doses of bisphenol A—though still trace amounts—produce effects not predicted by the toxicology models relied on by regulators. And this implies that the traditional style of risk assessments for endocrine-disrupting chemicals are inadequate for determining what every consumer cares most about: what constitutes a safe dose.

"If all you do is test high doses and then do some hand waving and divide by a safety factor, it's a good method if you don't want to find out you are wrong," vom Saal told me. "What the industry is hysterical about are the studies that show effects in animals at two hundred times below the levels that are in us. This basically says the whole regulatory system missed the safe exposure level of bisphenol A. And that means you cannot have bisphenol A in commerce. Because we know that if bisphenol A is used it will leach, if it leaches we will be exposed, and if we are exposed we will be harmed."

Those with a commercial interest in bisphenol A do not dispute that the substance leaches from packaging and containers that hold our food and drinks. Studies have detected it in soda pop and canned foods including tuna, peaches, pineapples, green beans, corn, soups, pasta, meal replacements, and infant formula. In a 2007 study spear-headed by the Environmental Working Group, one laboratory found it in more than half of ninety-seven cans and containers tested. University of Cincinnati researchers who simulated normal water bottle usage during back-packing, mountaineering, and other outdoor activities found that new and used polycarbonate containers released bisphenol A fifty-five times more rapidly after exposure to boiling water. This 2008 study suggests it's not the age of a container that matters most in determining how much it will leach. Instead, it's whether or not the container has been exposed to boiling water.

Another study determined that washing polycarbonate baby bottles in the dishwasher or with boiling water and a brush caused increased leaching. The chemical linkage is inherently unstable between bisphenol A molecules and the ester bonds that form a polymer used to make polycarbonate plastic. Age and repeated washings further destabilize the link and contribute to leaching. One study estimated that between birth and three months of age, babies who are bottle-fed formula and not exclusively breast-fed are exposed to the highest levels of bisphenol A relative to body weight. And in a 2007 test sponsored by the Environment California Research & Policy Center, five leading brands of baby bottles—Avent, Dr. Brown's, Evenflo, Gerber, and Playtex—leached bisphenol A at levels ranging from 5 to 10 parts per billion.

Exposures are higher from using liquid canned formula than from feeding from a polycarbonate bottle. Nevertheless, other polycarbonate containers used to store food and marketed for use in the microwave, such as Tupperware's Rock 'N Serv line, may increase leaching. These are

touted for their durability, suggesting you can use and
wash them over and over again, even though doing so increases the likelihood of leaching. And the problem isn't
confined to storage containers. A chemical analysis of paper towels and food take-out containers demonstrates that
they also contain bisphenol A. Paper towels made from recycled paper may seem like a good thing. Yet studies show
recycled product contains higher levels of bisphenol A
than that made with virgin material because of how
the chemical is used in paper production. Then there's
dentistry. Since the 1960s, several bisphenol A–derived
resin products have been used in preventative sealants, adhesives, and restorative materials. Studies of saliva show
that small quantities of bisphenol A leach from these dental materials for an hour or two following application. Finally, when products rooted in bisphenol A reach landfills,
they contribute to estrogenically active groundwater contamination. One study determined that while treatment of
landfill leaches removed 97 percent of estrogenic activity,
traces of bisphenol A remained.

As a mom and a scientist who researches bisphenol A,
Rebecca Roberts finds herself doing risk-benefit calculations when she goes to the store with her two-year-old
daughter. "Part of me is annoyed because bisphenol A
is so pervasive," said Roberts, an associate professor of
biology at Ursinus College in Collegeville, Pennsylvania.
"What do you do when your child wants the Dora the Ex-
plorer sippy cup made from polycarbonate plastic? I want
to make her happy, but as a scientist, I'm really concerned
about the low-dose effects of bisphenol A leaching from
polycarbonate plastic." Industry trade groups, however,
say the benefits of the chemical far outweigh any risks.

"Human exposure to bisphenol A from the use of protective liners for food cans is exceedingly low, hundreds to even a thousand or more times lower than safe exposure levels set by U.S. and international regulatory agencies," said Dr. William Hoyle, chairman of the Inter-Industry Group for Light Metal Packaging. As for polycarbonate plastics, it's the same story. "Daily intake is well below the standard set in January 2007 by the European Food Safety Authority," said Dr. Steve Hentges, executive director of the Polycarbonate/BPA Global Group of the American Chemistry Council. "Every authority that has looked at this has come to this same reassuring conclusion."

According to vom Saal and other frontline researchers, however, the safe exposure levels described by Hoyle and Hentges are absolutely meaningless. First of all, they fail to consider the scores of studies showing adverse effects at low doses. "While a billionth of a gram may seem like an incredibly small amount, bisphenol A in human cells causes effects at the low part-per-billion level," said vom Saal. Beyond that, biomonitoring studies of bisphenol A in human blood and tissue suggest that people already are being exposed to levels that far surpass the current government reference. Humans quickly metabolize bisphenol A. So in order to account for the levels detected, people must be exposed to ten times more than the current acceptable daily intake level. Indeed, the leaders of the House Committee on Energy and Commerce were so dismayed by the FDA's approval of bisphenol A for use in products intended for infants and children—which was based on two industry-funded studies—that they launched an investigation in April 2008.

The polymer scientists who synthesized bisphenol A

into plastic in the early 1950s did so knowing the substance mimicked estrogen. But it probably did not seem significant at the time. For one thing, it was years before polycarbonate plastics became an industry standard for food and beverage containers. For another, no one yet had the ability to measure the trace amounts of bisphenol A found in the polymers. However, pharmaceutical researchers looking for synthetic estrogens experimented with bisphenol A in the 1930s with a specific interest in the chemical's hormonelike effects. As a medication, it lost out to the more potent estrogenic drug diethylstilbestrol (DES), familiar to baby boomers and some Gen Xers because it was prescribed to pregnant women between the late 1940s and 1971, resulting in between five million and ten million exposures. (It was also used as a supplement in chicken and cattle feed for many years.) At the time of its use, DES was considered safe. But we now know that women who took DES to help prevent miscarriages and premature deliveries are at modestly increased risk for breast cancer. Their daughters have greater worries. They are at an increased risk for cancer of the vagina and cervix, structural abnormalities of the reproductive tract, pregnancy complications, and infertility, according to the CDC. Meanwhile, DES sons are at increased risk for noncancerous cysts of the epididymis (the tube inside the testes that stores and transports sperm).

One of the latest studies on bisphenol A suggests it may cause some of the same adverse effects as DES. Retha Newbold, a pioneering DES researcher at the NIEHS for more than thirty years, found that female mice exposed to low doses of bisphenol A while in the womb developed abnormalities in the ovaries and reproductive tract as they reached an age equivalent to midlife in humans. Among them: cysts inside and outside the ovaries, polyps, and excessive growth of the lining of the uterus. Newbold, who is among four researchers from U.S. agencies to sign the bisphenol A consensus statement, said her research suggests that bisphenol A could be associated with endometriosis and uterine fibroids, which are the leading causes of the six hundred thousand hysterectomies performed on U.S. women annually.

Newbold's study follows closely the publication of research from another laboratory that shows bisphenol A can cause chromosomal abnormalities in the grandchildren of pregnant mice dosed with the substance at levels within the range of common human exposures. Dr. Pat Hunt at Washington State University discovered that exposure to bisphenol A disrupts meiosis, or early egg development, in mouse fetuses. Like all mammals, female mice form their eggs while still in their mother's womb. Thus, not only is a fetus exposed, but so are eggs that will produce the next generation. After the animals exposed to bisphenol A in utero reached adulthood, Hunt's laboratory fertilized their eggs and found an increase in chromosomally abnormal embryos. "By hitting the mother with a low dose, we increased the likelihood the grandchildren would be abnormal," Hunt said.

This was Hunt's second major finding on bisphenol A. The first, in 2003, came about after an incident in the laboratory at Case Western Reserve University in Cleveland where she worked at the time. Hunt, a geneticist, had been conducting experiments with mice in order to get at why chromosomally abnormal pregnancies increase with age. All of a sudden her control colony showed an unexplainable increase in damaged oocytes, or developing eggs. Not only were there too few or two many chromosomes present, the chromosomes also were misaligned. After weeks of investigation, Hunt found the cause. A temporary lab employee had cleaned the colony's polycarbonate plastic cages and water bottles with a harsh detergent intended for use on the floor. The scrubbing caused the plastic to break down, which exposed the mice to bisphenol A at very low concentrations. Intrigued by the accidental discovery, Hunt decided to deliberately expose mice to small amounts of bisphenol A. Her results were similar to those discovered after the lab accident: the mouse eggs showed greatly increased rates of two particular chromosome abnormalities, which, Hunt pointed out, are the leading causes of miscarriage, congenital defects, and mental retardation in humans.

Immediately, the news zipped around the world that a chemical capable of leaching from food and beverage containers had damaged egg cells in mice exposed to very little of it. Could this be a problem for humans too? Though Hunt had been careful to say that she did not know what effects, if any, these exposures might have on people, she pointed out that biomonitoring data indicated that pregnant women are exposed to similarly low levels of bisphenol A. "Certainly we should be concerned enough to carry out extensive further study," Hunt said in a statement issued by NIEHS, which funded her work. What happened next was outside Hunt's experience as a published researcher: the chemical industry attacked her conclusions as "so speculative as to be scientifically dubious." The bisphenol A website (www.bisphenol-a.org), a function of the American Chemistry Council, criticized her for theorizing that an abnormal egg might produce an abnormal embryo.

Hunt was stunned by the industry's response. "They paid someone to read my paper and provide them with talking points," she said. "Damage control is very serious business to the chemical companies. They fight back."

No one knows this better than vom Saal, whose work on bisphenol A has been under attack from the moment he published his first study results more than a decade ago. For most of his career, vom Saal's research focused on blips in natural hormone levels that cause harm during critical periods of brain and reproductive tract development. In the mid-1990s, his lab was studying how the fetus is protected from the high levels of estrogen present during pregnancy. His research team discovered that a sex-hormone-binding globulin, essentially a barrier system in the blood, blocked natural estrogen from entering cells. They decided to see if this system worked similarly to block synthetic estrogenic chemicals. As it turned out, "The great majority of man-made chemicals are not inhibited from entering cells like natural estrogens are," vom Saal said. "They go right past this barrier system, and the receptor in the cell that causes changes when estrogen binds to it is, unfortunately, very responsive to bisphenol A."

In 1997, vom Saal's group reported that trace amounts of the chemical—2 parts per billion—fed to pregnant mice caused enlarged prostates in their offspring. Mincing no words about the implications of their findings, the researchers wrote that they "show for the first time that fetal exposure to environmentally relevant parts-per-billion doses of bisphenol A, in the range currently being consumed by people, can alter the adult reproductive system in mice." This immediately set the plastics industry to work funding a study of its own.

Predictably, the industry study could not replicate vom Saal's results. And for years, even as evidence of low-dose effects has steadily grown in the scientific literature, the industry behind bisphenol A has continued to repudiate low-dose findings. Relying on scientific reports written by product-defense specialists who, again, cut their teeth defending the tobacco industry, the American Chemistry Council dismisses 100 percent of the studies suggesting that bisphenol A causes harm at low doses. Steve Hentges of the ACC told me that the "weight of evidence" does not support low-dose findings. What that means, according to Hentges, is that all the research showing low-dose effects has been contradicted by more comprehensive studies that have not reproduced the findings. "There just isn't that much compelling evidence to suggest that bisphenol A exposure causes low-dose effects." After the city of San Francisco adopted a law banning the sale of baby products derived from bisphenol A, chemical and toy manufacturers used such arguments in a lawsuit against the city, and San Francisco wound up repealing the ban in May 2007, before it could take effect.

In reality, however, samples of scientific evidence showing how the chemical increases risks for an array of diseases now number in the hundreds. Over the past decade, studies have looked at behavior of the chemical in cell cultures or in animals. In a sobering 2008 study, researchers from the California Pacific Medical Center Research Institute and the Stanford Genome Technology Center found that low levels of bisphenol A caused normal, noncancerous human breast cells to express genes characteristic of aggressive breast cancer cells. "We screened 40,000 genes in normal human cells that had been exposed to BPA and found a striking increase in the sets of genes that promote cell division, increase cell metabolism, and increase resistance to drugs that usually kill cancer cells, and prevent cells from developing to their normal mature forms," said Dr. Shanaz Dairkee, principal investigator and coauthor of the report that appeared in the journal *Cancer Research*. "Breast cancer patients with this kind of gene expression tend to have a higher recurrence than other patients, and they have a worse survival rate." Other research has measured the levels of bisphenol A in humans and animals or analyzed the source of exposures. It's long been known that bisphenol A binds to estrogen receptors. But more recent studies suggest it is capable of provoking changes in tissue enzymes and interacting with other hormone-response systems, such as the androgen and thyroid hormone receptor signaling systems.

In sum, these studies suggest that low-dose bisphenol A exposure, given prenatally or to neonates, results in functional and structural changes in the prostate, breasts, testes, and mammary glands, and in the body size, brain structure and chemistry, and behavior of laboratory animals. It does so by altering epigenetic programming, or the expression of more than two hundred genes involved in how the brain gets wired, how cells multiply, and how metabolism is regulated. "The list of diseases and adverse health conditions now plausibly linked to bisphenol A by animal and cell research is large and reflects disease trends in the human population," wrote Pete Myers, in his eloquent essay "Good Genes Gone Bad." "It runs from reduced sperm count to spontaneous miscarriages; from prostate and breast cancers to degenerative brain diseases; from attention deficit disorders to obesity and insulin resistance, which links it to Type 2 diabetes." As Myers points out, bisphenol A is by no means the only synthetic chemical capable of altering gene behavior. "Some are compounds that have been of concern for decades, like dioxins, polychlorinated biphenyls (PCBs), and certain pesticides. Others, like phthalates, Teflon-related [perfluorinated] chemicals, and brominated flame retardants, have been attracting more attention recently, especially because they are present in consumer products in every home in America." But bisphenol A, because of its demonstrable effects at infinitesimal exposures, is at the forefront of the debate over the fetal origins of adult disease.

Loaded with a decade's worth of work linking bisphenol A to a list of life-altering health problems at exposures common to every one of us, more than three dozen of the world's experts on bisphenol A agreed it was time they did their own assessment of the state of the science. With funding from the National Institute of Environmental Health Sciences, the National Institutes of Health, the Environmental Protection Agency, and the nonprofit organization Commonweal, they examined more than seven hundred articles on bisphenol A with the aim of evaluating the strength of the data.

Five expert panels from different scientific disciplines were established prior to a meeting in Chapel Hill, North Carolina, in October 2006. Each panel began working on a draft report, which was posted on a conference website and distributed for all panelists to read. At the meeting, interdisciplinary groups integrated information from the reports and classified the data according to strength. Only when all groups agreed on a statement was it carried forward to the peer-reviewed published consensus statement written by vom Saal. In addition, all five preliminary reports were peer-reviewed and published by the journal *Reproductive Toxicology*. "This novel approach was like conducting an experiment with four replicates," said Dr. Jerry Heindel, the scientific program administrator for NIEHS. "It had never been done before for any environmental chemical." The consensus statement had political implications too. Not only were the thirty-eight signers putting their professional reputations on the line, they were setting up a scientific showdown with a panel assembled by the National Toxicology Program of the National Institutes of Health.

In 1998, the National Toxicology Program established the Center for the Evaluation of Risks to Human Reproduction as an environmental resource for the public and for regulatory and health agencies. As its name suggests, the center assesses synthetic and naturally occurring chemicals for their potential to adversely impact reproduction and development. To do so, the center convenes panels of scientific experts who review and evaluate the scientific data on a particular chemical. The panel then issues an opinion that is considered by the National Toxicology Program in writing monographs that characterize the potential for the chemical to cause reproductive and developmental harm. These detailed reports from the NTP carry considerable weight: they help set regulatory policies on chemicals at the state, federal, and international levels, said Dr. John Bucher, associate director. For example, the NTP's monograph on DEHP (diethylhexyl phthalate) was the basis for a July 2007 petition by physicians' groups, including the American Medical Association, asking the FDA to label medical devices that contain this phthalate. The NTP report stated, "There is serious concern that certain intensive medical treatments of male infants may result in DEHP exposure levels that affect development of the male reproductive tract."

The findings of the center's expert panel on bisphenol A would be factored into the final NTP report that would hold the same kind of sway. However, the panel's work was abruptly halted in March 2007 after the Environmental Working Group leaked to the *Los Angeles Times* that a consultant hired to coordinate the panel had a conflict of interest because of ties to the chemical industry. The NTP subsequently fired the consultant, Sciences International, whose client roster has included bisphenol A manufacturer Dow Chemical and five other chemical companies or groups. Sciences International was in the fourth year of a $5 million contract with the federal program. During that time it had helped prepare reports on seventeen chemicals. After firing the consulting firm, the NTP instituted conflict-of-interest guidelines, which it previously did not have. But the program let stand the first draft of a scientific review prepared by Sciences International on bisphenol A's risks after an audit concluded it included all relevant references and would be "useful" to the panel members. From this draft, the panel of nine scientists—chosen for their expertise in their various disciplines but only one of whom had direct experience researching bisphenol A—would draw their conclusions.

When it became public, the three-hundred-page draft report generated harsh criticisms from bisphenol A researchers. They called it biased in favor of the chemical industry, inaccurate, and tainted. "Is the panel purposefully misrepresenting data or just grossly misunderstanding it,"

wrote a group of exasperated bisphenol A researchers
from the Tufts University School of Medicine. In sixty
pages of painstaking detail, the Tufts research team, which
has produced groundbreaking work linking bisphenol A
exposure in rats to breast cancer and precancerous mammary tissue, excoriated the NTP draft, writing that "many
sections of the report illustrate a disregard for the nature
of science." The draft report excluded the Tufts research
from consideration by the NTP panel because bisphenol A was administered to test animals through a pump
rather than orally. Indeed, the report excluded dozens of
studies that were not orally administered. But, as the Tufts
researchers pointed out, "A fetus does not eat. It is exposed to bisphenol A through its mother's blood, and
studies of humans indicate that low levels of bisphenol A
are regularly detected in the blood." Dr. Marcel Maffini,
an assistant professor in anatomy and cellular biology at
Tufts, told me concern is so high regarding bisphenol A
because researchers are convinced it causes a syndrome in
lab animals "that parallels the DES syndrome. This chemical alters the endocrine system at very low doses. Our research is looking at the mammary gland, but others are
looking at brain development, behavioral changes, and
obesity. It takes a blind person not to see it."

The conflict-of-interest accusations swirling around the
NTP draft report combined with the unequivocally damning consensus statement released by the bisphenol A experts created the kind of controversy relished by the
media. Anticipating that the NTP panel might pronounce
bisphenol A to be as worrisome as described by the re-

search experts, national television, radio, and print reporters crowded into a hotel conference room in Arlington, Virginia, to hear the panel deliberate on August 8,
2007. Vom Saal, Maffini, and a handful of other bisphenol A researchers were there. Also present were some
industry-funded researchers and a public relations team
from the American Chemistry Council, including Steve
Hentges. The best-dressed among the crowd was Robert
Weiss, a class-action attorney who has filed a California
superior court suit against manufacturers of polycarbonate baby bottles and some grocery stores that sell them.

Inside the meeting room, the panel plodded through its
work. But just outside the room, the scene was theatrical.
Vom Saal, Hentges, and Weiss took turns before the cam-
eras and tape recorders, knowing they were competing for
that night's sound bites and tomorrow morning's quotes.
In the end, when the panel concluded it had "some concern" that bisphenol A exposure to fetuses, infants, and
children causes neural and behavioral effects, it wasn't the
kind of "gotcha" the journalists could make into a big
story. Nevertheless, it amounted to the first small step toward regulating the chemical. Though far from the sweeping indictment of bisphenol A issued just days earlier by
the thirty-eight scientists who signed the vom Saal–penned
consensus statement, the panel's conclusions were serious
enough for a warning from the panel chairman that it may
not be prudent to wait for scientific certainty to take protective action. "This might be a time for the application of
the precautionary principle," said Dr. Robert Chapin,
head of developmental toxicology at Pfizer. "If people
have concern, they can and should make choices that reflect those concerns. I think there is more of a need for cau-

tion for younger humans than for older humans." Not only did Chapin acknowledge that dealing with bisphenol A may require a shift away from traditional risk assessments, he gave parents good reason to consider alternatives to polycarbonate baby bottles.

The message was not lost on those who were listening. Sales of BornFree baby bottles, a line of bisphenol A–free products available online and in stores, nearly doubled in the days following news reports of the findings. The authors of the books *Baby Bargains* and *Baby 411* said they no longer recommended any of the most popular polycarbonate baby bottles on the market. And a Maryland state legislator promised to revive an effort to ban the sale of bisphenol A–containing products intended for children under the age of six. By early 2008, two large Canadian retailers, Mountain Equipment Co-op and Lululemon Athletica, stopped selling polycarbonate water bottles that leach bisphenol A. "These are retailers that get it," said Dr. Rick Smith, executive director of Environmental Defence, an advocacy organization. "I have no doubt that other large Canadian retailers will be following suit." Soon, they may have no choice. As this book was going to press, the Toronto *Globe and Mail* reported that regulators were on the verge of characterizing bisphenol A as a dangerous substance, potentially paving the way for actions to curb its use in Canada.

As for the United States, John Bucher, the associate director of the National Toxicology Program, told me at the time of the expert panel's decision that the NTP would review all the literature on the chemical and then draft a final report. Released April 15, 2008, the same day that word leaked of Canada's newfound toxic take on bisphenol A, the final NTP draft report agreed with the expert panel's conclusion that there is "some concern for neural and behavioral effects in fetuses, infants and children at current human exposure." But the NTP draft went further, pointing out some concern for exposure "based on effects in the prostate gland, mammary gland and an earlier age for puberty in females." After a scientific peer review, the NTP reduced its level of concern about bisphenol A's effects on the mammary gland and earlier puberty in girls. However, what the current science says about this ubiquitous chemical boils down to a few words in the final September 2008 NTP brief that U.S. regulators should weigh carefully—and without delay: "the possibility that bisphenol A may alter human development cannot be dismissed."

Fred vom Saal, the bisphenol A researcher who has refused to let the everyday hazards of this estrogenic compound go unnoticed, told me Canada's regulatory actions, the concerns raised by the NTP's brief, and a congressional investigation into the FDA's acceptable findings on bisphenol A created "a perfect storm" in the spring of 2008 that made U.S. action to limit exposures virtually inevitable. "It's clear to the public the scientific issues are no longer in dispute," said vom Saal. "The climate is such that if the FDA ignores science and makes a purely political decision [to maintain the status quo], it will lose all credibility." That is, the little the agency has left.

OUT OF THE FRYING PAN AND ONTO THE PAPER: PERFLUORINATED CHEMICALS

The bad news clung to Teflon like crud on a burned pan. A chemical used in making fluoropolymers, long-chain molecules that produce the nonstick qualities of the famous cookware, was in the blood of everybody in the United States, and it was surprisingly toxic to test animals. In 2002, the EPA launched a priority review of the substance—an industrial workhorse called perfluorooctanoic acid—and lawyers got busy building cases against DuPont, the chemical giant that makes Teflon. More than fifty thousand people living near a Teflon processing plant in West Virginia sued DuPont because their drinking water was tainted with PFOA. Nearly two dozen class-action lawsuits on behalf of Teflon-buying consumers flooded federal district courts. The EPA, in a sharply worded lawsuit, alleged that DuPont withheld evidence for twenty years that PFOA endangered workers and the public. And

a federal grand jury scrutinized DuPont's actions for possible criminal violation. Outside of the courts, concerns
about the safety of Teflon-coated cookware caused a panic
in China, where sales of nonstick products plummeted
and retailers removed some items from store shelves. The
EPA's scientific advisory panel advised classifying PFOA
as a likely human carcinogen. And California pondered
placing PFOA on its Proposition 65 list of chemicals that
cause cancer, birth defects, or other reproductive harm.

From his Wisconsin farm, the toxicologist Rich Purdy
monitors the ruckus regarding PFOA with keen interest.
It's a habit that's hard to break. Purdy's concerns about
the family of chemicals to which PFOA belongs dates to
the early 1980s, shortly after he joined 3M. At the time,
3M was the world's largest manufacturer of PFOA, which
it sold to DuPont and others. PFOA acts as a surfactant,
smoothing out the Teflon or nonstick coatings on cookware and other products. Scientists also refer to PFOA as
C8 because of its eight-carbon structure. In addition to its
role in producing Teflon, the chemical also works as a processing aid for other fluoropolymer applications, including
waterproof, breathable material (such as GORE-TEX),
plumbers' tape, and dental floss. Virtually every industry
uses fluoropolymer applications, according to the EPA,
including aerospace, automotive, construction, chemical
processing, electronics, semiconductors, and textiles. 3M
depended on PFOS for its Scotchgard brand of repellents
and paper packaging. It also sold it in bulk to outfits that
make carpets and textiles. Other uses for PFOS chemi-

cals included firefighting foams, hydraulic fluids, alkaline cleaners, floor polishes, photographic film, denture cleaners, shampoos, and insecticides.

In 1997, forty years after putting its perfluorinated chemicals in commerce, 3M was in the midst of a study of its workers when it discovered PFOS contamination in a control group of blood bank samples that were presumed to be clean. Investigating further, 3M scientists came to a shocking realization: the chemicals contaminated virtually everybody. Low levels of PFOS and PFOA were ubiquitous in the human population. The widespread exposure compelled the company to look closely at the toxicology of the substances. The data, some of it generated by Purdy, showed that PFOS was persistent, bioaccumulative, and toxic—the most troubling trio of properties to any toxicologist.

After much internal roiling, 3M abruptly announced to the public in May 2000 that it was reformulating Scotchgard and quitting the business lines associated with these chemicals—taking a $200 million charge against its bottom line. The decisions, 3M officials insisted in a press statement, were based on the company's "principles of responsible management" and did not mean the substances were unsafe. Carol Browner, the EPA administrator at the time, varnished 3M's announcement in a press statement noting that the company's actions would "ensure that future exposure to these chemicals will be eliminated, and public health and the environment will be protected." The move was applauded by the market and sent 3M's stock price up for the day. But Rich Purdy, who had left the company three months before the announcement, knew this was just the beginning of an environmental story about a family of chemicals that are so strong they never, ever break down.

At the Washington headquarters of the nonprofit Environmental Working Group, staffers were scratching their heads over the developments at 3M. Why would a major chemical company suddenly reformulate a wildly popular—indeed, iconic—product line at a tremendous cost to its bottom line? Seeking answers, the watchdog group asked the EPA for the entire scientific record on PFOS. "It's what they call the 'docket,' " said Richard Wiles, executive director. In this case, the docket consisted of "thirty thousand pages of science and correspondence back and forth on this issue." Reading through the record, Wiles and his colleagues got an eye-opening crash course in perfluorinated chemistry, and "it was really scary," said Wiles.

Not only were the chemicals ubiquitous in people, wildlife, and the environment, but weird, deadly things happened to test animals when 3M dosed them with PFOS. Records in the EPA docket show that cynomolgus monkeys exposed to relatively low levels exhibited low food consumption, excessive salivation, labored breathing, loss of coordination, significant reductions in serum cholesterol levels, and death. Rat studies by 3M were suggestive of developmental and reproductive toxicity, too. Indeed, on the day of 3M's greenwashed phaseout announcement, a high-ranking EPA official alerted colleagues around the world to the newfound problems with the chemicals, warning that there would be no quick fix.

"These chemicals are very persistent in the environment, have a strong tendency to accumulate in human and animal tissues and, based on recent information, could potentially pose a risk to human health and the environment over the long term," noted Charles Auer of the EPA's Office of Pollution Prevention and Toxics in an e-mail to other U.S. agencies and international regulators. Auer informed his colleagues that the "continued manufacture and use of PFOS represents an unacceptable technology that should be eliminated to protect human health and the environment from potentially severe long-term consequences." His memo also mentioned a fact 3M's public announcement did not: 3M was ending production of PFOA, leaving DuPont without a supplier for a processing aid it couldn't do without in order to make Teflon. Meanwhile, said Auer, the EPA was "examining its options regarding action on PFOA." Evidently, so was DuPont. After 3M made its last batch of PFOA in 2002, DuPont, which had not previously produced PFOA, took over from 3M as the principal manufacturer—and did so with full knowledge of the EPA's reservations about the substance. (DuPont declined through spokesman Dan Turner to make its representatives available for interviews for this book after learning I wanted to discuss, among other things, why DuPont decided to make PFOA after 3M got out of the business.)

DuPont already was in the business of manufacturing fluorotelomers. Sometimes referred to simply as telomers, these small perfluorinated polymers competed against 3M's PFOS and have the same grime-fighting and grease-protecting qualities. Telomers also present many of the same environmental and public-health challenges. The big difference between the competing chemical products came down to manufacturing: 3M used an electrofluorination process to make its now-discontinued PFOS family of chemicals; DuPont and other manufacturers of fluorotelomers continue to employ a different process known as telomerization.

If you visit DuPont's Teflon website, you can take a cyber tour of what the company calls Teflon World. Click on the bedroom scene and you'll learn that Teflon products—some fluoropolymers and others fluorotelomers—impart strength to nail polish and stain-resistance to carpets, drapes, upholstery, clothing, linens, dog beds, and luggage. DuPont is the world leader in fluorotelomers through its Teflon and Zonyl brand names.

During 2004, an estimated 72,000 tons of products were sold containing 11,250 to 13,500 tons of active fluorotelomer ingredients, according to the Organisation for Economic Co-operation and Development. Like PFOS-related products, fluorotelomers fight grime and stains on clothing, carpeting, and soft furnishings. They repel oils and greases on food containers such as fast-food wrapping, pizza boxes, microwave popcorn bags, paper plates and cups, and the liner boards for candy bars and bakery items. They're integral to household and commercial products such as polishes, waxes, and window cleaners. Paints, printing inks, copy paper, and lubricants make use of them, too. Not a hint of PFOA is used in manufacturing fluorotelomers. But PFOA is an unintended by-product, or residual, of the production process. In some fluorotelomer products, PFOA is found at trace levels. Significantly, data suggests that Mother Nature breaks down fluorotelomers to form PFOA in the environment, where it does not degrade any further. And in a wrinkle that sounds as if it could be ripped straight from the pages of sci fi, once inside living organisms fluorotelomers appear to metabolize into PFOA through a process that some scientists suspect involves chemical intermediates that could be orders of magnitude more toxic than PFOA itself.

As troubling as the record appeared on perfluorinated substances, the staff at the Environmental Working Group quickly saw an opportunity: put the issue before the public and press for change. Formed in 1993 as a nonprofit research and public-policy organization, EWG uses government data, legal documents, scientific studies, and its own laboratory tests to counter corporate spin in policy discussions and news stories. Since 2004, an arm of EWG has lobbied Capitol Hill. Funded by more than two dozen private charitable foundations with politically progressive agendas, EWG is the Prius-driving tree-hugger set's answer to big-business-oriented think tanks. Some of its detractors have dubbed it the Environmental Worrying Group, and they dismiss its work as unduly alarmist and unscientific. Regardless of what one thinks of the group, it has evolved into a powerful player on issues ranging from agriculture to energy to toxics. From its loft-style headquarters in Washington, D.C.'s U district, EWG's media-savvy staffers churn out the jargon-free reports and web-accessible databases that make it a favorite of journalists and broadcasters working on deadline. Its first report on perfluorinated chemicals made news when it tied these substances to popular brand names such as Scotchgard, GORE-TEX, and Teflon.

Later, according to Richard Wiles, when DuPont was sued by neighbors of a Teflon processing plant in West

Virginia because of PFOA-contaminated water, EWG worked through plaintiffs' lawyers to obtain DuPont documents from the litigation. In those documents, EWG found evidence that, twenty years earlier, DuPont had detected PFOA in the umbilical cord blood of pregnant plant workers, two of whom delivered babies with birth defects. But DuPont did nothing to inform the EPA. Based on the information, the group petitioned the EPA "to bring charges against DuPont for violating the one component of federal toxics law that has any teeth, which is a requirement to inform the agency when there are adverse effects caused by your chemical," Wiles said. After investigating, the EPA agreed with EWG that DuPont had failed to meet its obligation in that incident and in others involving community water testing and blood serum sampling around the West Virginia plant. The agency, in a settlement agreement, ultimately tagged DuPont for a $16.5 million penalty for violations of the Toxic Substances Control Act, representing the largest administrative fine ever levied by the EPA. Of the amount, some $5 million was earmarked for independent testing programs that are currently assessing the potential for fluorotelomers to degrade to PFOA in the environment. Other testing, also paid for by the settlement, is under way to assess the potential for fluoropolymer and fluorotelomer products to release PFOA as they age while in use, potentially leading to exposures.

For decades, scientists and regulators assumed perfluorinated chemicals were not toxic and, because they were incorporated in polymers, biologically inert. This was incorrect. Instead, the substances represent a new type of persistent organic pollutant that, well, redefines persistence. PFOA, PFOS, and closely related substances are virtually indestructible. They owe their muscle to a carbon-fluorine bond that is among the strongest in organic chemistry. "PFOA and other perfluorocarboxylic acids are about as unreactive as any chemicals we know," said Scott Mabury, a chemistry professor at the University of Toronto who is one of the world's leading researchers on these substances. "You take DDT, the thing that Rachel Carson was concerned about, and PFOA makes it look ephemeral. DDT has a half-life in soil of ten years. Put these in soil and you will never see them degrade under any conditions. You simply can't predict what the half-life would be."

That perfluorinated chemicals last forever in the environment—coupled with the fact that they do not, in organisms, take to fat like other persistant pollutants but instead bind to blood proteins and blood-rich areas such as the liver—makes them difficult to understand. Toni Krasnic, a chemist who coordinates the EPA's perfluorinated chemical activities, described them to me this way: "These are the most peculiar chemicals we've seen in probably the entire study of chemicals." Traditional risk assessment models are not sufficient in characterizing these chemicals, said Krasnic, which makes the work of EPA scientists very slow going. To understand the properties, sources, and pathways of exposure requires new research that can take years to complete.

Lacking statutory authority to effectively regulate chemicals it suspects of being bad actors, the EPA has taken steps attempting to limit future releases of the controversial perfluorinated chemicals. The agency negotiated a voluntary agreement with manufacturers calling for three things: a 95 percent reduction in the facility emissions of

PFOA and chemicals that break down into it by 2010; a 95 percent reduction in the use of PFOA and related chemicals in consumer products by 2010; and a commitment from manufacturers to "work toward eliminating these chemicals from emissions and product content no later than 2015." Separately, the EPA issued a new rule to restrict the return to the U.S. market of 88 PFOS-related chemicals phased out by 3M. (Three exemptions were made for highly specialized, low-volume uses, including airplane hydraulics, semiconductor manufacturing, and photographic imaging.) Finally, announcing it could "no longer conclude that these polymers will not present an unreasonable risk to human health and the environment," the agency proposed a rule, still under consideration as of this writing, requiring manufacturers to enter a previously exempted review process for all PFOS- and PFOA-related polymers.

Given the persistent, bioaccumulative, and toxic nature of long-chain perfluorinated substances, the leaders of the United Steelworkers, which represents workers at DuPont and pulp and paper mills, believe the EPA's actions are insufficient to protect workers and the public from their hazards. The union, part of a coalition pushing California to list PFOA as a cancer-causing chemical under the state's Proposition 65 act, has alerted thousands of manufacturers and retailers that they may have a legal duty to inform consumers of potential PFOA hazards related to their products. Companies that were sent such notices include Rug Doctor, Stanley Steemer, McDonald's, Taco Bell, Papa John's, Pizza Hut, KFC, California Pizza Kitchen, Levi Strauss, ConAgra Foods, Gap, W. L. Gore & Associates, Eddie Bauer, J.Crew, Wal-Mart, Sears, Nordstrom,

Dillard's, and Dalton Carpet Outlet. "While the EPA has finally recognized the obvious, it must create more pressure on the manufacturers of PFOA to eliminate it altogether and find a safe substitute," said Leo Gerard, USW president. "We urge all companies that sell products containing PFOA—directly or as breakdown product—to provide warnings to their customers." Ed Schneider, a spokesman for W. R. Gore & Associates, which relies on fluoropolymers to make GORE-TEX, said his company weeds out pollutants before they can reach consumers. Schneider told me the pressure is on the chemical industry to come up with alternatives free of trace contaminants. "We would like to have raw materials that don't contain it. We don't need it or want it, and it costs us effort and equipment to eliminate it," Schneider said.

For its part, DuPont was the first of eight companies to sign on to the voluntary PFOA phaseout program agreed to by manufacturers and the EPA. (The others are 3M/Dyneon, Arkema Inc., AGC Chemicals/Asahi Glass, Ciba Specialty Chemicals, Clariant Corp., Daikin, and Solvay Solexis.) In February 2007, DuPont announced it had met the reduction benchmarks for 2010—three years ahead of schedule. Moreover, Chairman Charles O. Holliday, Jr., said the company was "committing to end its need to make, buy or use PFOA by 2015." Yet the news did nothing to assuage DuPont Shareholders for Fair Value, who, for the third year in a row, demanded a report on PFOA and potential alternatives at DuPont's 2007 annual meeting. As in the past, the shareholder proposal failed, but it poked at a sensitive area for the company by spotlighting DuPont's potential liabilities and revenue losses associated with PFOA concerns. "The demand from retailers and other manufacturers is such that we don't think they will wait until 2015 not to grab up alternatives if DuPont continues to use PFOA," said Sanford Lewis, a lawyer for the share-holders group, which includes the United Steelworkers.

As of early 2009, the EPA was continuing its risk assessment on PFOA. It will eventually set an "acceptable level of risk" for human exposure. The agency has yet to act on the recommendation made by its scientific advisory board to list PFOA as a likely human carcinogen, which would trigger a full EPA risk assessment for cancer. In the meantime, California legislators passed a ban on food packaging that contains more than 10 parts per billion of perfluoro-chemicals. Senate Bill 1313, which would have prohibited the manufacture, sale, or distribution of PFC-tainted food packaging beginning in 2010, came under fire by industry lobbyists, and Governor Arnold Schwarzenegger vetoed it in September 2008. The paucity of information and standards—coupled with the slow pace of independent research—leaves plenty of room for industry to character-ize the continued use of PFOA and chemicals that degrade to it as completely benign. As part of a class-action settle-ment, DuPont is funding community-based health screen-ing near its Parkersburg, West Virginia, plant. The results won't be known for years. In the meantime, DuPont spent $1 million on peer-reviewed testing that concluded Teflon brand products are safe to use. And in 2006, the company announced that results from a health study of thousands of workers and former workers at the West Virginia plant showed there are no known human health effects associ-ated with exposure to PFOA. But the reality is far more complicated and less rosy. The EPA, FDA, and regulators around the globe are actively investigating PFOA and related chemicals—and concerns about hazards are many.

The Society of Toxicologists held a special meeting in Washington in early 2007 so researchers could share the state of the science under way in government, industry, and academic programs. Researchers are keenly interested in how perfluorinated chemicals might be negatively impacting the liver, the immune system, reproduction, and development. Much attention focused on new evidence connecting high levels of PFOA in baby rodents with obesity later in life. Though research is preliminary, it's another example of why toxicologists worry that children are an especially vulnerable population. Some scientists think PFOA's effect on fatty acid metabolism is interfering with the programming of energy metabolism and creating an undernourished environment in utero. This fits with what is known about human fetal programming syndrome. Children who have it were chronically short of nutrients in the womb. As these children grow up with abundant food, they are more likely to become overweight because their metabolism wasn't programmed properly during development. The concern is that PFOA exposure at a critical period of development might affect metabolism for a lifetime.

As for DuPont's worker health study, members of the company's own Epidemiological Review Board, a team of independent experts from such prestigious institutions as Johns Hopkins and Yale universities, disapproved of the way DuPont characterized the results, describing it as "somewhere between misleading and disingenuous," according to an article by Ken Ward, Jr., of *The Charleston*

Gazette. Another e-mail from members of the expert board urged DuPont to present a more accurate picture of the study's findings. "We believe that no party can claim sufficient knowledge that PFOA does or does not pose any risk to health," the board members wrote on March 2, 2006, to DuPont officials. "Thus, we question the basis of DuPont's expression asserting that PFOA does not pose a health risk." The e-mails were made public as part of a federal court filing in a lawsuit over PFOA pollution from a DuPont plant in New Jersey.

DuPont's actions all seem to fit with a strategy outlined in a proposal DuPont received from a consulting group seeking to help the company deal with its regulatory and legal problems regarding PFOA. Unearthed in the EPA docket by a reporter for *Environmental Science & Technology*, the proposal by the Weinberg Group stated: "The constant theme which permeates our recommendations on the issues faced by DuPont is that DUPONT MUST SHAPE THE DEBATE AT ALL LEVELS."

Against this backdrop, it's not surprising that DuPont is quick to point out that Teflon is not PFOA. Indeed, Teflon itself is a substance known as polytetrafluoroethylene, or PTFE, which was first synthesized, serendipitiously, by a DuPont chemist named Roy J. Plunkett. One of Plunkett's compressed, frozen samples of tetrafluoroethylene, from which he hoped to make a refrigerant, turned from a gas into a white solid. Somehow, the small molecules of tetrafluoroethylene had spontaneously reacted to form a polymer, or a giant molecule. In this molecule, hydrogen atoms were replaced by fluorine atoms. The result was a substance that was resistant not only to water but also to fats and oils. Plunkett eventually devised a way to reproduce the polymerization process in the laboratory, though DuPont did not know what to do with the chemistry until the early 1940s, when the military put it to use in the atomic-bomb-building Manhattan Project. DuPont changed PTFE's mouthful of a name to the much catchier Teflon. Cookware coated with Teflon, described by DuPont as "a housewife's best friend," hit the market in the 1960s. Since then, billions of Teflon pots and pans have been sold in some forty countries.

Pop culture endowed the Teflon name with new meaning, one synonymous with politicians and personalities who always seem to slither out of trouble. In large part, the Teflon brand has done just that. California, as of this writing, has denied requests to place PFOA on its Proposition 65 list of cancer-causing chemicals. And in October 2007, the U.S. Justice Department ended its criminal investigation of DuPont without any charges. True, the bad news about PFOA hit parent company DuPont where it hurts: in a $108 million settlement of the West Virginia class-action suit (with another $235 million on the line if an independent science panel determines a link between PFOA and human disease); in a record $16.5 million administrative fine paid to the EPA; and in the specter of even larger payouts if the company fails to successfully defend itself in twenty-two consumer class-action lawsuits pending as of this writing. Whatever these costs amount to, they may represent a true bargain for DuPont. The company believes $1 billion of its annual $27 billion in revenues are at stake if the EPA or another agency were to regulate or prohibit PFOA, according to a February 2008 filing with the Securities and Exchange Commission.

Ever since the world woke up to problems with perfluorinated chemicals, DuPont has let Teflon-covered cookware take the heat. And the pots and pans have managed to stay in the kitchen. Sales of nonstick cookware increased 5 percent in 2006, according to the Cookware Manufacturers Association. More significant, the focus on Teflon and other nonstick cookware—which experts agree is really not the meat of the perfluorochemical problem—has drawn attention away from stain- and grease-repelling fluorotelomers and their potentially pivotal role in PFOA pollution and exposure. It's a kind of bait-and-switch that, so far, has saved DuPont a lot of trouble over the thornier issue with fluorotelomers that has scientists and regulators on alert: namely, that fluorotelomer polymers—even after being stripped of problematic impurities as planned by DuPont and other manufacturers—may break down to form telomer alcohols that then break down to form PFOA. As of early 2008, research was ongoing in this area. "That is the big outstanding issue," the EPA's Krasnic told me. "Do the polymers themselves break down?" If the answer is yes, then the problem is endemic to the structure of flurotelomer polymers, which are widely used across our economy. Because of such concerns, Canada recently prohibited four new fluorotelomer-based substances that break down to form PFOA and related chemicals.

At the heart of this whole mess are the studies showing that perfluorinated chemicals have dirtied the planet and everyone on it over the course of the past half century. Since 2000, these synthetic chemicals have been detected in human blood samples in the United States, Japan, Canada, Colombia, Brazil, Italy, Poland, Germany, Belgium, Sweden, India, Malaysia, Korea, China, and Australia. More recently, researchers have looked for and found them in breast milk, liver samples, seminal plasma, and umbilical cord blood.

Wildlife is tainted, too—from mink to field mice, from bottle-nosed dolphins to ringed seals, from Brandt's cormorant to polar bears. The highest concentrations are found in the livers of fish-eating animals living closer to industrialized areas. In the environment, there is simply nowhere on Earth you can escape it. PFOA and other perfluorinated chemical pollutants have been detected in the Tennessee River downstream from a fluorochemical manufacturing plant. They are in drinking water sources near production plants in West Virginia and Germany. And they are in places you would never expect to find them. The Arctic is polluted. So are the coastal waters of South China, Japan, and Korea, tributaries to China's Yangtze and Pearl Rivers. They fall with rain. They mix with the air we breathe and the dust in our homes. They taint soil, sediment, wastewater, and sludge worldwide.

Scientists and regulators are keenly interested in sorting out how perfluorinated chemicals get around. No one is busier with this work than Scott Mabury. A bearded, balding researcher who chairs the chemistry department at the University of Toronto, Mabury has been described as the Wayne Gretzky of fluorinated compounds for his prolific research in the area of environmental fate. His research team turns out study after study, and Mabury frequently finds himself sitting before regulatory agencies or on the phone with journalists. His work is cutting-edge and essential. Until scientists pinpoint how the contamination occurs, it will be hard to find solutions.

That perfluorinated pollutants can be measured in areas where perfluorinated chemicals are made and used is not so hard to understand. Emissions occur during manufacturing. They happen when consumers clean and care for products treated with the chemicals, sending contaminants into wastewater treatment systems. The chemicals then enter the aquatic environment and get into the food web. Perfluorinated pollutants also leach from landfills containing discarded consumer goods. In addition, starting materials that remain in finished products can escape during use and then break down in the environment to form perfluorinated pollutants.

But what has puzzled people like Mabury is how these substances, at relatively high levels, get to remote and pristine environments such as the Arctic. On this front, two theories have emerged. The first describes a legacy of point-source releases by industry, of which there are plenty, and hinges on long-range transport of pollutants via oceanic currents. Until it was hit by a lawsuit in the late 1990s, DuPont, for example, sent tens of thousands of pounds of PFOA emissions flowing into the Ohio River. These pollutants traveled our waterways and hitched rides to the Arctic through oceanic currents. So, in the case of the DuPont plant at Parkersburg, West Virginia, PFOA flowed down the Ohio, joined the Mississippi River, emptied into the Gulf of Mexico, and then traveled up to the Atlantic to finally reach the Arctic.

The other hypothesis points to an indirect source involving highly volatile leftover starting materials—alcohols used in making fluorotelomers—that are released from manufacturing processes and the products themselves. As Mabury explained to a Canadian Senate committee in early 2007, "We have hypothesized that they act as the travel agents for getting the basic structural unit to the Arctic. Then in the atmosphere they degrade into the [perfluorinated] acids that we measure in seals and polar bears. They move up the food chain."

Are there enough starting materials, or residuals, around to cause this effect? For many years, according to Mabury, most of the makers of perfluorinated products insisted they did not leave any residuals in their finished products. In theory, the production processes they used removed all the volatile fluorinated alcohols, none of which have functional benefits. "It was fallacious," Mabury told me. "No chemical synthesis is 100 percent efficient." The best chemist in the world cannot get a 100 percent yield on a reaction. That means there will always be impurities as a production byproduct. Mabury's team, in fact, has measured residuals in industrial and consumer products. On a weight basis, they amount to a few percent. It may not sound like a lot, but 2 percent adds up to about 250,000 kilograms a year, or just over 550,000 pounds. Mabury's team also has measured concentrations of perfluorinated pollutants all across North America. Maintaining the measured concentrations would require residual releases of about 250,000 kilograms a year. This contributes to Mabury's confidence that residuals are involved.

The research team calls its hypothesis Precursor Alcohol Atmospheric Reaction and Transport, or PAART, and team members have produced more than forty papers reporting various measurements that test the theory. In one experiment conducted in April 2006, Mabury sent a graduate student to the Devon Ice Cap in the Canadian Arctic to sample how much of these pollutants come in every year from the atmosphere and how this was changing over time. The student, a hearty woman named Cora Young, braved the Arctic in old-school cold-weather gear in order to avoid contaminating her experiment. Modern GORE-TEX gear, of course, is made with fluoropolymers to impart the water resistance that keeps adventurers warm and dry. Though she did not stay as snugly in her gear as she would have liked, Young brought back valuable measurements of the flux, a fancy term for how much of a substance is delivered per square centimeter in mass per year.

In total, the evidence collected by Mabury's team is consistent with PAART being the major source of Arctic contamination. It all adds up, said Mabury. "These impurities—starting materials—that are brought along that serve no function in the material used by the consumer may be a significant or potentially even the most significant source of these compounds to the environment." Much of what Mabury's team has found contradicts the theory of direct releases from production sources being responsible for the Arctic's fluorinated fugitives. "It's probable, eventually," Mabury told me. "But it takes a long time to swim to the Arctic. I'm skeptical if any significant quantities have made it."

Of course, scientists and regulators would also like to understand how perfluorinated substances get into all of us. Generally speaking, experts don't like how perfluorinated chemicals hang around in humans for a long, long time. The estimated biological half-life for PFOA is 3.8 years. For PFOS, it's 5.4 years. That means, for example, a body will rid itself of half of its PFOA in 3.8 years, half of what's left of that in another 3.8 years, and so on. "Anything with a half-life in years is no good," explained Christopher Lau, lead research biologist with the EPA's national Health and Environmental Effects Research Laboratory, where the agency is conducting research on PFOA and other closely related chemical pollutants. Adding to the concern is our constant exposure to PFOA through everyday activities, Lau told me. Biomonitoring studies show the average level of PFOA in humans is about 5 parts per billion per kilogram of body weight.

In 2005, the FDA began investigating consumer products that contact food as potential sources of human PFOA contamination. In addition to cookware, a research team analyzed food wrappers and papers treated with grease-repelling fluorotelomer coatings. "The FDA assumed three things," said Mabury. "First, that the fluorinated surfactants on the food packaging would not move into the food. Second, that if it did move into food, it would not be bioavailable—that is, it would not enter the bloodstream. And third, that if it was bioavailable, it would not actually metabolize and release the alcohols that break down into PFOA. It appears that all three assumptions were incorrect."

In fact, what FDA researchers found may make you think carefully about nuking that next bag of popcorn. Microwave popcorn bags contain a higher level of chemical coating than in any other product, and they get extremely hot extremely fast. This may be why the concentrations of chemicals migrating from the bags into the popcorn oil were hundreds of times higher than the amount of PFOA from heated nonstick cookware. Indeed, based on the concentrations measured by the scientists, consumption of 10 bags of microwave popcorn a year could contribute about 20 percent of the average PFOA blood level, according to a report by *Environmental Science & Technology*'s Online News. It's not the pans we should worry about: it's the paper.

In 2007, Tim Begley, an FDA scientist, reported experiment results suggesting a perfluorinated substance called polyfluoroalkyl phosphate surfactant (PAPS) on food wrappings can migrate into emulsions such as butter, margarine, and chocolate spread at unexpectedly high rates, which he measured in the parts-per-million range. More than a dozen perfluorinated chemicals are FDA-approved for treating paper or other materials that contact food. But none has been officially evaluated by the agency for contact with foods that use emulsions or emulsifiers, according to an investigation by *ES&T*'s Online News. The FDA experiment dovetailed with work from Mabury's lab in 2007 demonstrating that rats dosed with PAPS metabolized it to form PFOA. As of this writing, his team was investigating whether the intermediate chemicals formed as the substance metabolizes are, as Mabury suspects, more toxic than PFOA.

Consumer groups concerned about public health and toxics have hit retailers and food companies hard over the concentrations of perfluorinated pollutants the FDA found in microwave popcorn and food containers. "We'd love to see these big chains make the phaseout of PFOA in food packaging be the first order of business," said Sandy Buchanan, executive director of Ohio Citizen Action. The group has coordinated the delivery of more than twenty-three thousand handwritten letters to grocery store managers and challenged ConAgra Foods, the maker of Orville Redenbacher popcorn and dozens of other food products, to disclose its use of perfluorinated chemicals. The effort led to a 2006 shareholder resolution requesting options for ConAgra to reduce or eliminate the use of PFOA-related chemicals in product packaging. In response, ConAgra said it would begin high-priority studies for various alternatives to eliminate fluorocarbons from food contact in its popcorn packaging.

As a practical matter, the FDA has not issued warnings about any perfluorinated products. But that didn't stop Burger King from phasing out fluorotelomers in 2002, with McDonald's following soon after. The EPA, as of this writing, said it "does not believe there is any reason for consumers to stop using any products because of concerns about PFOA." But the EPA also is quite clear that what it knows about PFOA gives it pause. "The science is still coming in, but the concern is there," said Susan Hazen, the acting assistant administrator of the EPA's Office of Prevention, Pesticide and Toxic Substances, in early 2006, when the EPA went public with its plea to industry to curtail PFOA production. "Acting now is the right thing to do for our health and the environment."

Unfortunately, there's new evidence suggesting the damage is already done. In August 2007, two studies appeared in the journal *Environmental Health Perspectives* that showed that babies exposed in the womb to perfluorinated chemicals may be born slightly smaller than other infants. It's preliminary work that must be duplicated, but the studies from Johns Hopkins University in Baltimore and the International Epidemiology Institute (which received funding from 3M) were the first to show an association between a negative effect and everyday levels of PFOA in people.

In test animals, higher doses of perfluorinated chemicals than those measured in humans caused enlarged livers, thyroid and immune problems, neonatal deaths, delays in growth and maturation, and testicular, liver, and pancreatic cancers. For some chemicals there is a narrow margin between a dose that causes mild effects and a dose that causes severe effects. Researchers are investigating the reasons and are zeroing in on several suspected mechanisms of toxicity that may or may not apply to humans. One obstacle is that humans and laboratory animals eliminate the substances quite differently. There are gender differences, too. In terms of half-life, it ranges from hours for the female rat to days for the male rat to months for a monkey to almost four years in humans. Without an understanding of the biological events that account for the differences, it's difficult for toxicologists to extrapolate results from one species to another.

The former 3M toxicologist Rich Purdy is concerned that the risk assessments that hold sway over critical policy decisions do not adequately consider the impacts on toddlers and infants, who are most vulnerable to perfluorinated substances. One reason is excessive exposure— babies are dosed in the womb, through their mother's milk, and when they suck on their sleepwear or crawl on the carpet, Purdy told a Canadian Senate committee reviewing the effectiveness of the country's Environmental Protection Act. The second reason is timing. "This group is exposed at a susceptible time in life. In that first year, we develop our immune responses." Referring to a pivotal 3M monkey study, Purdy said, "All the animals in the monkey study at the low dose had shrunken thymus glands. You and I don't care about our thymus glands. We are too old. The gland starts large when we are young, when we need it, because during that time the immune responses are set. If someone has the wrong ratio of thymus cells at that time, they will be prone to asthma." They also might be prone to early-onset diabetes because the thymus makes a mistake and does not destroy the right antibodies. If Purdy had his way, fluorinated chemicals would be assessed cumulatively. "With cumulative risk assessment, we look at all the chemicals in the class and identify chemicals that look alike. All the fluorochemicals look alike. They are acids with a bunch of fluorines on them. They have the same mechanism of action . . . the tissues attacked are the same." Not only would this be more efficient than assessing each substance individually, he said, it would be more protective.

Sad as it seems, it's not really surprising the EPA would know so little about the potential hazards of PFOA and closely related chemicals more than fifty years after chemists created them. In 1938, when DuPont's Roy J. Plunkett stumbled onto fluoropolymer chemistry, the U.S. EPA was still forty years away from existence. Then, in the 1970s, when the EPA and regulations governing toxic substances were created, industry lobbyists saw to it that more than sixty thousand compounds already in commerce would be exempted from safety testing—including perfluorinated substances. In a perfect world, our ingenuity would never create unintended consequences. But far too often, that's not the way it works. Plunkett himself created the first fluoropolymer while trying to synthesize a new refrigerant from chlorofluorocarbons, or CFCs. At the time, no one knew that CFCs would deplete the ozone, and CFCs represented a huge improvement over earlier refrigerants that were highly toxic.

Whether current concentrations of the perfluorinated chemicals in human blood will be shown to be significant in compromising human health is beside the point, said Mabury. "Nobody wants these things in their blood in those kinds of concentrations. In my view, they should not have to be there. Chemists should be good enough chemical architects to design materials that provide desirable properties without adverse pollution problems."

3M went back to the laboratory and developed a new Scotchgard line based on a chemical called perfluorobutane sulfonate or PFBS. DuPont, as of this writing, has not unveiled alternative chemistry for its fluorotelomers. Studies conducted by 3M and reviewed by the EPA indicate that 3M's new chemical is better behaved than the old PFOS chemistry and the fluorotelomers still made by DuPont. PFBS is based on a four-carbon chain that has a low potential to bioaccumulate. Currently, work is under way at the National Toxicology Program to determine how carbon chain length affects toxicity and persistence. Shorter chains are believed to be preferable in both regards. At the same time, there is other encouraging news to report: PFOS already appears to be declining in the Arctic based on measurements taken by Mabury's team just three years after 3M ceased manufacturing it. "We were surprised by how fast we saw a reduction," said Mabury. Furthermore, a 2007 study by the CDC suggests blood levels of PFOS, PFOA, and related chemicals have

declined since 3M's phaseout of PFOS, although more
data is needed to characterize this possible trend. That's
the good news; on the flip side, the same work shows the
chemicals contaminate 98 percent of the population, confirming ubiquitous exposure. And in Minnesota, the feel-
good news about 3M's switch to alternative chemistry is
largely overshadowed by the toxic legacy of its old perfluorinated chemical line. During the production of PFC
products between the late 1940s and 2002, wastes and
discharges from company operations fouled the local
environment, contaminating stream water, groundwater,
drinking water, soil, and fish. As of early 2008, 3M was defending four lawsuits related to its historical disposal of
PFOS and PFOA. Separately, the company has already
struck multi-million-dollar agreements with government
entities to help clean up the messes it left behind. "It is
true that PFOA and PFOS are still detected in the environment at low levels and that they do not break down so
it is important to look at ways of reducing that footprint
and certainly stop making future contributions," said
company spokesman Bill Nelson. "That is what 3M is doing and has done since 2000."

Rich Purdy, the former 3M toxicologist who left the
company because, he said, it did not act fast enough on
the PFOS issue when red flags emerged, has tempered his
view of his former employer over the years, noting it's
pretty easy to be blinded by self-interest. "The way I see it,
3M ran a stoplight, kept going for a little bit, and then
stopped," he told me. As for DuPont, "They ran a stoplight, hit somebody, and just kept going."

8 REACHING AHEAD: NEW POLICIES

On December 13, 2006, the European Parliament approved landmark legislation affecting thirty thousand chemicals used in the making of products as common as televisions, computers, cosmetics, furniture, and food wrapping. Though the sweeping reforms contained in the REACH legislation were years in the making, they boil down to one very simple concept: safety. Under REACH, chemical companies doing business in the European Union will lose access to its nearly five hundred million consumers and an $11 trillion market unless they can prove their substances are harmless. Strange as it seems, tens of thousands of chemicals that were developed before the enactment of environmental protection laws in the United States, Canada, and the European Union had been exempted from assessments to determine their risks. The European Union and Canada for decades had operated under chemicals policies modeled after U.S. rules. But policy makers outside the United States were dissatisfied with the results. They saw no sense in continuing to let loose untested chemicals, then chase down the ones that created messes in people or places.

A week before the historic REACH vote, Canada announced a new national chemicals management plan that, like the new EU legislation, is far more stringent than anything in the United States. After categorizing twenty-three thousand substances based on their potential to harm human health and the environment, Canada's regulators zeroed in on five hundred they determined to be the most dangerous (including phthalates, bisphenol A, and perfluorinated chemicals). Canadian officials warned that they may demand substitutes after conducting more detailed safety assessments. But the real pressure point is in the European Union, where the burden of proving that commonly used chemicals are safe has flip-flopped from public authorities to chemical manufacturers and importers.

As befitting the stakes, REACH was not a bargain made quickly or quietly. Lobbying leading up to the vote was, by all accounts, the most brash and bitter that Brussels had ever experienced. The €556 billion EU chemical industry, which is the world's largest producer, importer, and exporter of chemicals, spent freely attempting to persuade lawmakers to drop the effort or dilute it. On the other side, environmental and consumer groups, with the support of the public-health community and sympathetic EU officials, waged an aggressive public education campaign on the hidden dangers of toxic substances. Drama was high. Volunteers ranging from EU commissioners to multiple generations of families agreed to have their blood screened for the toxic substances that, as the by-products of modern living, pollute every one of us. (In a quirky political commercial, four leading members of the European Parliament meet during a "chemical therapy group" and confess they are contaminated with toxic chemicals. "I couldn't help it," says Riitta Myller, a Finnish member from the European People's Party, as others in the commercial scrunch their faces in distate. "I just bought a new carpet.") On the other side of the coin, the U.S. government got involved, and in an unprecedented manner: the Environmental Protection Agency, the Office of the U.S. Trade Representative, and the departments of state and commerce all lobbied against the proposal at the behest of the U.S. chemical industry.

An eleventh-hour compromise left all parties to the REACH negotiations with something to complain about. Nevertheless, the legislation, eight years in the making and light-years beyond U.S. policy requirements, signified a radical move toward better protections while avoiding undue costs to industry. According to European Parliament President Josep Borrell of Spain, REACH "sets up an essential piece of legislation to protect public health and the environment from the risks of chemical substances, without threatening European competitiveness. It offers EU citizens true protection against the multitude of toxic substances in everyday life in Europe."

Back in the United States, the EPA's Office of Pollution Prevention and Toxics, which implements U.S. chemicals policy, was whistling in the dark. OPPT hosted a conference in Austin, Texas, on the day of the long-scheduled EU Parliament vote. But the EPA gathered chemical industry representatives, environmental and public-health workers, and state officials from across the nation *not* to discuss the sweeping implications of the EU's chemicals policy reform. Strangely enough, the EPA's agenda for the Texas event had been carefully constructed to avoid focusing on REACH. Instead, the EPA was championing a program of its own called the High Production Volume (HPV) Challenge. Under it, companies—after avoiding doing so for nearly three decades—are making public rudimentary toxicity data on about two thousand eight hundred chemicals produced or imported in amounts exceeding one million pounds a year. Compared to the heft of REACH, the HPV program is a boy among men. But it's all the EPA can muster.

The chemical industry, for all intents, had been embarrassed into accepting the HPV Challenge. In 1976, when Congress enacted the Toxic Substances Control Act, it deemed that manufacturers should develop adequate data on chemical substances and mixtures. But it did not mandate the production of that data or give the EPA the proper tools to compel it. So in 1997, when the nonprofit advocacy organization Environmental Defense checked the public record, it found that basic data was missing for 71 percent of the best-selling chemicals in the United States. This meant the EPA had not been able to perform preliminary screening, let alone the more extensive testing that determines if a chemical causes specific types of harm. Indeed, when the EPA did its own investigation in 1998, it found that only 7 percent of high-production-volume chemicals had a full set of information illustrating their properties and effects. Scientists call this type of data "endpoints," and they include information about acute toxicity and repeat-dose toxicity, developmental and reproductive toxicity, mutagenicity (the ability to alter genes), ecotoxicity, and environmental fate. In its report, aptly titled *Toxic Ignorance*, Environmental Defense called for immediate action. "Chemical safety can't be based on faith," the authors wrote. "It requires facts. Government policy and government regulation have been so ineffective in making progress against the chemical ignorance problem, for so long, that the chemical manufacturing industry itself must now take direct responsibility for solving it. It is high time for the facts to be delivered." A short while later, in 1998, the HPV Challenge was born.

If EPA officials had wanted to avoid comparisons between the HPV program and the new REACH legislation in the European Union, their timing for the Texas conference couldn't have been worse. In the nine years the EPA has been running the HPV program, it has produced slow and sketchy results. As of mid-2007, the industry had not bothered to sponsor or submit data on more than two hundred HPV substances, forcing the seriously constrained EPA to issue rulings if it is to compel the submissions. Meanwhile, data sets received by the EPA were incomplete for nearly half of all chemicals in the HPV program. In contrast, during a similar period, EU politicians, industry leaders, government officials, and activists hammered out a complete overhaul of a complicated chemicals management system that will fill information gaps on the hazards of substances and identify appropriate risk management measures to ensure their safe use. The legislation, effective in June 2007, requires the registration of thirty thousand substances produced or imported in volumes over 1 metric ton (2,240 pounds) with a new Helsinki-based European Chemicals Agency. The most hazardous of these chemicals, an estimated fifteen hundred, could be banned or restricted. Chemicals of highest concern or those used in the greatest volume must be registered first, and all chemicals are to be registered by the time REACH has been in effect for eleven years. Under REACH, the European Union's industrial sectors are expected to vault ahead of the United States in developing safer, cleaner technologies and products, turning the United States from a leader into a laggard. And where human and environmental health are concerned, EU regulators now have a lever with which to extract crucial health and safety information from chemical manufacturers (no data, in effect, means no registration and no access to the market). Meanwhile, the United States, operating under the outmoded and ineffectual Toxic Substances Control Act, has no practical regulatory system for assessing the hazards of chemicals and controlling those of greatest concern. Instead, it must rely on voluntary programs, such as the HPV Challenge, and negotiated agreements with makers of suspect substances such as brominated flame retardants and perfluorinated chemicals.

"REACH is the world's most ambitious attempt to eliminate the dangers of untested, unregulated chemicals that are found at work, in our homes, and in our bodies," said Daryl Ditz, senior policy adviser at the Center for International Environmental Law in Washington, D.C., on the day the European Parliament passed the historic REACH agreement. "To protect the competitiveness of U.S. companies, we must now overhaul our own laws on toxic chemicals." Ditz, who served on the steering committee for the EPA conference in Austin, was not in attendance. He decided to stay home, he told me, after officials from the Office of Pollution Prevention and Toxics and representatives of the U.S. chemical industry quashed efforts to include an expert panel on the new European regulatory plan.

It wasn't supposed to be this way. As the Toxic Substances Control Act was being written in the early 1970s, its framers envisioned it as the crown jewel of the environmental policies to be administered by the newly created EPA. The legislation was revolutionary in that it was intended to prevent problems with chemical pollutants—not just manage them after they'd been created. Chemical companies were required to notify the EPA at least ninety days before manufacturing a new substance and provide data that would allow the EPA to assess risks. The idea was to give the EPA a chance to identify and regulate the worst chemicals before they ever got into the pipeline. Over the years, the EPA's reviews of new chemicals have resulted in some action being taken to reduce the risks of about 11 percent of the thirty-two thousand new chemicals that manufacturers have submitted. However, as mentioned, TSCA does not require chemical companies to test new chemicals for toxicity or to gauge exposure levels. And most chemical companies do not voluntarily perform such testing. So without this data, the EPA relies on scientific models to predict the hazards of new chemicals. Unfortunately, the system provides only limited assurance that health and environmental risks are being identified, according to a 2005 report by the U.S. Government Accountability Office.

As for the sixty-two thousand chemicals already in

commercial circulation when TSCA was enacted, the legislation was supposed to give regulators the authority to collect the data needed for evaluating their potential hazards. "We know so little—so abysmally little—about these chemicals," said EPA administrator Russell E. Train in October 1976. "We know little about their health effects, especially over the long term at low levels of exposure. We know little about how many humans are exposed, and how and to what degree." TSCA was intended to close this information gap. But more than thirty years later, Train's description of the problem is chillingly spot-on. Since the inception of the law, the EPA has used its authority to require testing for fewer than two hundred of those sixty-two thousand chemicals.

So what went wrong? When I reached Train, now in his mid-eighties, by telephone in Florida and asked him this question, he told me there was never much political will behind toxics legislation. "President Nixon recommended a Toxic Substances Control Act as part of his environmental message in the early 1970s," said Train, who went on to be president and chairman of the World Wildlife Fund before his retirement. "But not much happened, other than the chemical industry was very much opposed to the whole thing and Congress wasn't exactly excited about picking it up. I finally made a speech about it at the National Press Club, which was very unlike me, because it sort of waved a red flag about the dangers of chemicals. I tried to excite the public a bit. And that's when things finally got going."

Train, however, didn't recall much about the sausage making that went into TSCA. For that kind of background, he referred me to Terry Davies, the political scientist who drafted the legislation while he was a senior staffer at the Office of Environmental Quality, which preceded the formation of the EPA. I expected Davies, now a senior fellow at a Washington think tank called Resources for the Future, to have an interesting perspective about the limitations of the legislation. But what he said floored me. "I think TSCA at this point has proven to be a totally inadequate act, a totally toothless act that's about as close to worthless as you can get." The legislation, Davies explained, gave the EPA broad authority to act in order to protect human health and the environment. But the agency's ability to exercise its authority was effectively blocked by requirements, explored earlier in Chapter 2, spelling out how the law must be implemented. In effect, the EPA is in an impossible position because it lacks the power to request data on a chemical *prior* to proving it causes harm.

Back in Austin, high-ranking officials praised the HPV program as an example of a successful partnership with industry. Jim Gulliford, a silver-haired EPA veteran who was appointed assistant administrator for pesticides and toxics in time to defend TSCA before the U.S. Senate Committee on Environment and Public Works in the summer of 2006—kicked off the conference with remarks that echoed his earlier testimony on Capitol Hill. "We have complemented TSCA's tools and authority with the HPV Challenge," said Gulliford, who is on the record with Congress in his belief that TSCA provides the EPA with everything it needs to adequately review and control chemicals. Predictably, the chemical industry also gave the HPV program good marks. "By golly, an awful lot of information is in, and for a voluntary program that's remarkable," said the American Chemistry Council lobbyist and spokesman Steve Russell. He credited the HPV program with giving EPA "actual data on which to base priorities." And in a roundabout way, he acknowledged the limitations of TSCA by praising the voluntary program for making more information on more chemicals available to the EPA faster than any regulatory program ever could.

But as the conference unfolded over three days in the Radisson Hotel's windowless meeting rooms, it soon became clear that the HPV program was far less satisfying to those on the front lines of protecting public health and the environment from hazardous substances. The state workers and public-interest advocates trying to use the HPV data to assess the hazards of commonly used chemicals and evaluate potential alternatives expressed deep frustrations and myriad complaints about its utility, its completeness, and even its integrity. The EPA is just beginning to review HPV submissions for quality and completeness and to develop preliminary hazard assessments for chemicals considered to be of highest concern. Reviews of lower-priority chemicals won't be completed until 2010. And then there's the problem of more than two hundred "orphan" chemicals for which no information has been received at all. In short, the best effort the United States has so far mustered seems impossibly feeble—especially in comparison to the new European Union model.

"Our federal chemicals policy gives us the lowest common denominator to work with," said Bruce Jennings, a senior adviser to the California Senate Environmental Quality Committee, during one work session. "What we need is a more comprehensive approach that gets us away from dealing with the chemical problem du jour. We simply cannot do this one chemical at a time." As similar sentiments spilled over in other conference work sessions, industry folks became defensive. "This is insulting," shouted James Cooper, a senior manager and lobbyist for the Synthetic Organic Chemical Manufacturers Association, which represents 275 specialty chemical manufacturers. "This is an HPV conference. Let's just stay on topic, please." By this point, however, the elephant in the room could not be ignored.

As Terry Davies explained to me, protecting the status quo with respect to TSCA is important to both the industry and the EPA. "It's what I call their Faustian bargain: Industry likes TSCA because they don't have to do anything, and the EPA likes it because they don't have the resources to do anything else." The Office of Pollution Prevention and Toxics, which administers TSCA, operates with only 350 people and, like the rest of the EPA, an ever-shrinking budget under the Bush administration. Between 2004 and 2008, the EPA's budget decreased by 16.6 percent to $7.2 billion. In Senate testimony during the summer of 2006 before the Committee on Environment and Public Works, Dr. Lynn Goldman, a former EPA assistant administrator, cited budgetary constraints as among the reasons for the legislation's failure to live up to expectations. But Goldman, a pediatrician and professor of environmental health sciences at Johns Hopkins University, told the senators of other concerns. Chief among them is what she called "the enormous gap" forming between the requirements of the Toxic Substances Control Act and the EU's REACH legislation.

Before REACH, chemicals policies in the European Union and the United States had similar limitations. The vast majority of the chemicals in the EU lacked health and safety data because they were placed on the market prior to 1981, the year the European Community enacted its previous chemicals legislation. REACH, however, erases the artificial distinction between "old" and "new" chemicals and requires the registration of all chemicals produced or imported in volumes exceeding one metric ton. Chemical manufacturers and importers must share health and safety information with companies that use the substances in the production of consumer products, as well as with the public. And a few thousand of the most hazardous chemicals will require formal authorization, which is expected to provide a strong incentive to substitute safer alternatives.

Equally significant, REACH sidesteps the slippery slope of risk-benefit balancing, which doomed the EPA's ban on asbestos. Substances considered to be very persistent and those that accumulate up the food chain are not allowed if substitutes are available. If a substitute cannot be immediately identified, chemical makers and importers will be required to come up with a plan to find one. "Safer chemicals. Better understanding and management of chemicals through the supply chain. Improved worker protection. These are the things the European Union will get from REACH," said Dr. Joel Tickner, project director of the Lowell Center for Sustainable Production and assistant professor in the Department of Community Health and Sustainability at the University of Massachusetts Lowell. "It's a huge cultural change, certainly beyond our wildest dreams in the United States."

At least at the federal level. But Tickner and other experts see a rising tide of innovations by state and local governments, picking up the slack left by the void in national policy. State efforts, which tend to be concentrated in the West, Northeast, and Great Lakes regions, include right-to-know initiatives such as labeling, toxics-use data collection, and biomonitoring; procurement policies based on green chemistry or the precautionary principle; toxics-use reduction plans; and finally, restrictions on certain chemicals. Indeed, state officials who attended the Texas conference gathered early one morning to share information and discuss how to better coordinate their efforts and, ultimately, prepare for a push for new federal legislation. Said Ken Geiser, codirector of the Lowell Center for Sustainable Production, "We are issuing a historical invitation to the people working at the state level, asking them to begin to explore precursors to what we are going to do at the national level when the opportunity presents itself." Many of the states are looking to California, which is already rumbling with plans for its own REACH-like chemicals policy.

The man behind a California chemicals proposal is a soft-spoken assistant research scientist who spent fourteen years as a paramedic and firefighter before turning his attention to graduate school and environmental health sciences. As a doctoral student at the University of California at Berkeley, Michael Wilson investigated occurrences of peripheral neuropathy in automotive technicians who had high-level exposures to n-hexane in brake-cleaning solvents. The workers experienced nerve damage that caused

potentially permanent numbness, tingling, and weakness
in their feet, legs, hands, and arms. Through his investigation, Wilson learned how devastating a lack of data can
be: the manufacturers of brake-cleaning solvents used
n-hexane as a substitute ingredient for perchloroethylene
after California banned perc and all other chlorinated
compounds because they are probable human carcinogens. In effect, the effort to regulate one problem chemical
led manufacturers to substitute a different chemical without understanding its toxicity. And this, Wilson told me,
illustrates both the inefficiency and peril of our current
chemicals policy. "The chemical-by-chemical approach to
regulating substances is a model we have to get rid of," he
said.

After receiving his doctorate in 2003, Wilson was hired
by the Centers for Occupational and Environmental
Health. Soon after, he began work as the principal author
of a report requested by a state senate committee interested in exploring how the nation's largest state might deal
differently with chemicals. By 2050, California's population is expected to climb by about 50 percent to fifty-five
million. By the same token, global chemical production is
expected to grow about 3 percent a year, or double every
twenty-five years. Wilson and his coauthors argue that a
chemicals policy is a key element in California's transition
to a sustainable future. "Problems associated with society's
current approach to chemical design, use and management represent one of the major challenges of the 21st
century, and reorienting this approach will require a
long-term commitment to the development of a modern,
comprehensive chemicals policy. In California, chemical

problems are already affecting public and environmental health, business, industry and government. On the current trajectory, these problems will broaden and deepen."

The report generated a heady buzz in California, where policy makers, industry leaders, and the public view the world through green-colored lenses. Under Governor Arnold Schwarzenegger, California vaulted in front on the climate change issue with legislation that mandates greenhouse gas reductions by major industries. Wilson's report makes a compelling case for the state to similarly jump ahead on chemicals policy reforms. In it, he identifies three problem areas associated with the failures of U.S. chemicals policy. The first one is "the data gap," the lack of comprehensive toxicity information because chemical producers have not been required to generate and disclose it. The second is the "safety gap," the difficulty of controlling hazardous chemicals because of the lack of data and ineffectual regulatory tools. And the third is the "technology gap," the creation of market conditions that favor existing chemicals and defer the development of cleaner, safer and sustainable products.

To close these gaps and craft a revolutionary chemicals policy for California, the report proposes instituting a chemical reporting system, expanding regulatory authority, and creating market incentives and government support for the development of green chemistry, also known as clean or environmentally benign chemistry. In sum, Wilson told me, the ideal California chemicals policy would shift the chemicals market to make less hazardous materials as important to manufacturers and their customers as the function, price, and performance of a chemical substance. "Once that happens," said Wilson, "it will drive the research and development of greener products."

Green chemistry, which promotes pollution prevention at the molecular level by designing benign alternatives to hazardous chemicals and processes, was introduced in the 1990s by two brilliant young U.S. chemists, John Warner and Paul Anastas. "The dream is to have molecules so safe in the first place that regulations are not so important," said Warner. "But that's a dream that is generations off." Just getting started requires a fundamental shift in how chemists approach their work. In their book *Green Chemistry: Theory and Practice*, Warner and Anastas write, "Much like the Hippocratic procedures and protocols, a synthetic chemical methodology, to be truly elegant, must 'first, do no harm.' " This means chemists must consider the ultimate effect on human health and the environment when designing new substances. And it is quite a challenge for chemists, most of whom were not required to study toxicology in their doctoral programs.

Anastas, a professor at Yale University who previously ran the Green Chemistry Institute at the American Chemical Society, told me that unleashing the power and potential of green chemistry requires significant investments in research, development, and implementation, efforts that Congress has so far failed to adequately fund. But Anastas, who coined the term "green chemistry," said the name is as much about the color of money as it is about cleaner chemicals. "People think I'm joking about this, but I'm not. For generations now, we've thought that environmental benefits had to be a cost drain. That if you wanted something clean, it was going to cost you. And if you wanted to make money you were going to have to dirty up the place a little bit. Green chemistry breaks the fallacy. You can make things cleaner, more environmentally benign, and have them be more profitable. People say, 'Oh, this will save you money?' The answer is: No, this will make you money."

The Presidential Green Challenge Awards, presented annually since 1996 by the EPA, illustrate some promising examples. SC Johnson, the maker of home cleaning and storage products, is reformulating many of its offerings using a system that rates the environmental footprint of its ingredients. Doing so helped the company switch to low-density polyethylene in its popular Saran Wrap, eliminating the use of nearly four million pounds of polyvinylidene chloride a year. Similarly, Cargill Dow, now known as Cargill, developed a product that uses corn instead of petroleum to make a plasticlike film wrap. The product, called NatureWorks PLA, received a huge boost when Wal-Mart began using it to package fresh-cut fruit, herbs, strawberries, and brussels sprouts at Sam's Club and Wal-Mart Super Center stores. Because of the volume involved, replacing the conventional packaging on just four items saves the equivalent of eight hundred thousand gallons of gasoline and reduces more than eleven million pounds of greenhouse gas emissions a year, according to Wal-Mart.

Businesses intent on scouring hazardous substances out of their supply chain have set a green wave in motion by demanding product innovations and safer alternatives. Catholic Healthcare West and its forty hospitals gave Braun Medical a five-year, $70 million contract for medical bags and tubes that are free of polyvinyl chloride

(PVC) and phthalates. Meanwhile, Kaiser Permanente, the largest private health-care provider in the United States and the largest private-sector employer in the San Francisco Bay Area, pressured carpet manufacturers to come up with PVC-free products for use in all its new facilities. "There is enough evidence about the hazards of vinyl that the responsible thing for a health-care company to do is replace it," said Lynn Garske, Kaiser's environmental stewardship manager. Mercury-free thermometers, latex-free gloves, and recyclable cleaning products are among Kaiser's earlier successes under its preferable purchasing plan.

And the trend is not limited to health-care providers. A 2006 report by Clean Production Action identified Herman Miller furniture, InterfaceFABRIC, Avalon Organics cosmetics, Dell computers, and the specialty clothing retailer H&M as leaders in adopting business strategies that are healthier to humans and the environment. (Indeed H&M demonstrated its commitment to a progressive chemicals policy a few years ago by pulling a highly promoted Christmas item. Ads showing famous models posing in H&M underwear were already in circulation when H&M learned the sequins used to decorate some of the undies contained polyvinyl chloride. H&M scuttled the skivvies at considerable cost.) For each of these companies, the first steps in building sustainable practices involved assessing the hazards of chemicals used in their supply chains, then demanding that their suppliers substitute the most dangerous ones or rework their production processes to avoid creating them.

As U.S. businesses reevaluate their materials and processes and consumers become more aware of chemical dangers, some of the most progressive states are seeking to capitalize on the demand for safer alternatives. Michigan governor Jennifer Granholm issued an executive directive calling for state agencies to determine ways to give incentives to businesses and universities that pursue green chemistry programs. Massachusetts is expected to launch a similar economic development strategy based on green chemistry. Meanwhile, a future is fast approaching in which fossil fuel reliance will be impossible and waste-generating processes won't be tolerated. A committee of experts formed by the National Academy of Sciences has warned the chemical industry that it has less than twenty years left for the use of fossil fuels as its predominant source of energy and chemical feedstocks. After 2025, the health of the industry rests with its ability to carry out green chemistry and engineering, which is built on the fundamental understanding of the full life cycle impacts and toxicology of chemicals.

Yet a challenged U.S. chemical industry—saddled with a trade deficit, higher natural gas costs than its global competitors, and huge price increases in nonrenewable fossil fuels—is sidestepping investments in research and development needed for new technologies. Talking is cheaper: back in 1996, the U.S.-based Council for Chemical Research together with all major U.S. chemical industry organizations established ambitious goals for a shift to new technologies that incorporate environmentally and economically safer processes, use less energy, and produce no harmful by-products. Even today, the fifty largest U.S. chemical companies all state they are committed to sustainability goals. But their spending on research and development has decreased or remained flat since about 2000, when the chemical industry entered a down cycle unlike any it has experienced since the Great Depression. As a result, the authors of the California chemicals policy proposal argue that the time has come for policies that "support, motivate and compel" the industry to shift to new technologies, including green chemistry.

"Technology transitions can occur reactively in response to a loss of market share—as the experience of the U.S. auto industry illustrates—and they can be spurred proactively through public policy," they write. "Industry leaders recognize that technology transitions are inevitable, and, in fact, are the driving force of innovation and new growth . . . On the current trajectory, however, most U.S. chemical companies will likely continue to rely on existing chemical technologies and products, and some of these companies will end up in a reactive transition as they attempt to remain solvent in an increasingly competitive global economy."

Investments in current technology—and a friendly and familiar regulatory framework—create considerable resistance. And to be fair, nobody really likes change. Just like people, businesses often need a catalyst in order to do things differently. In the European Union, the catalyst was REACH. Parliament members deflected arguments that billions in costs and millions of jobs lost would stymie most research and development of newer, cleaner substances and processes. In the end, the vote over this hugely ambitious undertaking wasn't even close. The European Parliament decided by a wide margin that the benefits of REACH far exceeded the costs. "REACH will not be free," wrote European Commission vice president Margot Wallström on her official blog, "but the alternative [paying for damage to health and the environment] is even more costly." With the approval of REACH, the U.S. status quo becomes harder to defend. Like it or not, REACH signifies a powerful new business reality that has already crept into the United States, said Daryl Ditz of the Center for International Environmental Law. "The downstream users of chemicals—the car companies, the electronics makers, the producers of cleaning solvents—are demanding safer alternatives. You can have a thousand Greenpeace people screaming about 'Chemical X,' but when General Motors says they're not going to buy it anymore, lots of people jump."

In early 2008, Wilson and the Centers for Occupational and Environmental Health issued a second report on chemicals policy with instructions for California policy makers on exactly how to close the data, safety, and technology gaps that stand in the way of consumer safety and product innovation. It was signed by 127 faculty of the University of California. That's two dozen more than the number of faculty who signed the university system's report on climate change that led to California's mandatory greenhouse gas reductions. Green chemistry underpins the proposed reforms. "California's ability to link economic opportunity with human health and environmental protection will be a cornerstone for a sustainable future," the authors wrote.

Looking ahead, said Wilson, "The chemicals policy question in California is not whether it will come to be, but what is it going to look like?" The answer to that question began taking shape in September 2008, when Governor Schwarzenegger signed legislation authorizing the California Department of Toxic Substances Control to identify chemicals of concern, evaluate alternatives, specify regulatory responses, and provide the public with access to toxicity information. The six-hundred-member Chemical Industry Council of California supported the laws, as did Environmental Defense. But as of early 2009, much work remained to ensure that the groundbreaking California toxics laws—intended to make human health a priority—did not fall short of that imperative. Indeed, the California laws, unless strengthened and clarified through further legislation and executive action, will *delay* restrictions on well-known hazards such as bisphenol A and perfluorinated chemicals because of newly required reviews and bureaucratic red tape. One of the biggest problems, said Ditz, is that the California laws fail to compel companies to provide hazard data, unlike the European Union's REACH, which shifts the proof-of-safety burden from regulators to chemical manufacturers. To an industry accustomed to writing its own rules for the past thirty years, a requirement for basic toxicity information as a condition of market access may seem radical. To the rest of us, it's just common sense.

EPILOGUE:
MY LIST AND BEYOND

When it comes to toxic exposures, the personal is political. Curtailing exposures to hazardous chemicals in consumer products requires individual and collective action. It would be easy to shrug your shoulders and continue to buy and use the same things. But small adjustments in your own life can lessen your exposures and attendant risks. In the appendices that follow, you'll find suggestions for how to get started. Because I am frequently asked, "What changes have you made to reduce the type of chemical exposures you write about?" I wanted to share my own list. Here, in no particular order, is what I've done to lighten my chemical load:

- I buy and eat organic foods, whenever possible, because they are pesticide-free.

- I swore off microwave popcorn because of the fluorotelomers contained in the packaging.
- I ditched all my plastic food containers because they can leach hazardous substances when heated in the microwave. I use glass or ceramic containers instead.
- I canceled my contract for monthly bug control inside and outside my home because it's really not necessary.
- I decline all optional stain protection treatments for upholstery or floor coverings that merchants try to sell me.
- I use low VOC paint for my home-improvement projects.
- I replaced my bisphenol A–derived polycarbonate Nalgene bottle with a Sigg bottle made from aluminum.
- I filter my tap water.
- I vacuum and dust my home and office at least once a week (which, I admit, is more than I did before I wrote my book) because dust is loaded with the chemical pollutants that concern me.
- I bought hard-anodized aluminum pots and pans and no longer use nonstick cookware.
- I ask retailers questions about the things I buy. If they don't know the answers, I contact manufacturers.
- I read labels. Sometimes they don't give me the entire picture, but at least they can provide me with clues.
- I talk to my friends and family about the changes I've made and why I've made them.

Really, that's it. I don't obsess about chemical pollutants; I make informed decisions based on my understanding of the hazards of pesticides, plasticizers, flame retardants, grease repellants, and stain protectors. When I know something contains suspect substances, I ask myself: Can I find an alternative?

The answer is usually yes, though in one case, it took a while to find an appropriate alternative. I work in an old building where water the color of a caffe latte runs from the tap. Filtering doesn't make it any more palatable. So the best way for me to get the copious amounts of water I drink every day is to dispense it from a five-gallon jug perched atop a rented cooler. For years, the only type of jug available was made from bisphenol A–derived polycarbonate plastic. Every couple of months or so, I would check with various water delivery companies to see if they could supply bottled water in glass or a different type of plastic. (I also used the opportunity to explain my concern about polycarbonate and bisphenol A.) It took two years until a bottled water company in my area finally switched to plastic containers made from polyethelyne teraphthalate (commonly known as PET or PETE), and I gladly transferred my business.

If I had been serving a child, or at a time in my life when I was considering having a baby, I would have stopped using polycarbonate water jugs even before finding an alternative. But that wasn't the case, so I waited to make the switch while I pressed for change. Based on my particular circumstances, I decided the benefit of having drinking water in my office outweighed the risk posed by bisphenol A leaching into that water. Unfortunately, all consumers are faced with making similar calculations because our regulatory system fails to assure that even the most commonly used chemicals are safe as currently used.

By sharing this list, I do not mean to suggest that we can

shop our way clear of hazardous substances. We can't. Large-scale, population-level exposure reductions will happen only when we eliminate the worst-acting substances in favor of safer alternatives. To that end, Congress must reform the Toxic Substances Control Act with provisions demanding proof of safety from manufacturers. The European Union has already done it, and we can, too. What's lacking, so far, is the political will.

Happily, this is about to change. The Obama administration promises to make reform of our nation's easily manipulated and tragically broken toxics laws a top priority of the Environmental Protection Agency. Indeed, Lisa Jackson, in a memo to her staff after being sworn in as agency administrator in January 2009, pledged "that I will not compromise the integrity of EPA's experts in order to advance a preference for a particular regulatory outcome." Jackson stated emphatically that managing chemical risks would be among her top priorities—on par with reducing greenhouse gas emissions, improving air quality, cleaning up hazardous waste sites, and protecting the nation's water. "More than 30 years after Congress enacted the Toxic Substances Control Act," she wrote, "it is clear that we are not doing an adequate job of assessing and managing the risks of chemicals in consumer products, the workplace and the environment. It is now time to revise and strengthen EPA's chemicals management and risk assessment programs."

A starting point might well be legislation known as the Kid-Safe Chemicals Act. This common-sense proposal would use biomonitoring to detect chemicals in people and make these substances the subject of priority safety reviews. Furthermore, chemicals detected in human umbilical cord blood, which is indicative of the toxic exposures babies receive before birth, would be presumed unsafe and placed at the top of the safety review list. In addition, the Kid-Safe Chemicals Act would make newly required health and safety data from manufacturers available to the public in a centralized online database. If both consumers and institutional buyers become wiser, more informed shoppers, it can't help but put pressure on manufacturers to innovate safer substances (as we saw with the rush to find an alternative to bisphenol A–derived polycarbonate baby bottles).

State and local governments have made the moment ripe for federal action by moving ahead with restrictions on individual substances. Even industry concedes that federal rules are less burdensome than a patchwork of restrictions that vary by jurisdiction. Most important, sometime in 2009, the CDC is releasing its next biomonitoring study, which reports on U.S. population exposures to some 250 chemicals, close to twice as many as the 2005 report. The much-anticipated scorecard builds the case for what we already know:

We are the body toxic, and we can no longer afford our ignorance.

Appendix 1:
IT'S ALL ABOUT YOU

Hundreds of chemicals are in you, in me, in all of us. That much we know for sure. But at this point, we don't know exactly how much the hangover from this chemical cocktail may be harming us. In the meantime, reducing exposures can't hurt and is an especially good move in households with young children, pregnant women, and women who expect to become pregnant. Below you'll find some tips you can use to reduce your own exposure to the chemicals discussed in *The Body Toxic*. If you're appalled (and you should be) that the Toxic Substances Control Act does not provide the information needed to protect you and your family from the hazards of commonly used chemicals, tell your representatives at the federal level. You can find them at www.congress.org/congressorg/home. But don't let your local and state elected officials off the hook, either. Many of the most innovative and sweeping changes to make our world safer are being instituted at city halls and in state capitols.

CHEMICAL: ATRAZINE

Atrazine is the most commonly applied agricultural pesticide in the United States, where its heaviest use is on corn and sorghum crops in the Midwest and Southeast. It's also among the most common herbicides found in ground- and stream water, according to the United States Geological Survey. The EPA considers atrazine safe in drinking water at concentrations of less than 3 parts per billion. But the work of Dr. Tyrone Hayes of the University of California at Berkeley shows it causes gonadal abnormalities in amphibians at concentrations that are thirty times lower than what the EPA considers safe.

Are you exposed?
You can ingest small amounts of atrazine in drinking water, particularly in the spring in areas where the pesticide is heavily used. Farmworkers are exposed to higher levels of atrazine and other pesticides and bring it into their homes on their clothing and shoes. Studies in the Midwest have detected low levels of atrazine in carpet and house dust.

What to do?
To weed out atrazine and other pollutants, filter your tap water. Some carbon-based water filters, commonly available in grocery stores as pitchers or faucet mounts, can remove atrazine from drinking water. Check the product

labels. To search for water quality reports from your area, go to the EPA's Office of Water at www.epa.gov/water. Remove your shoes to avoid tracking any pesticide contaminants into your home. Vacuum, mop, and dust all surfaces weekly. For more resources, contact the National Pesticide Information Center at 800-858-7378 or npic.orst.edu.

CHEMICAL: PHTHALATES

Phthalates, which give plastic flexibility and resilience, are found in many consumer items, including personal care products, detergents, and soaps, and in the things made from polyvinyl chloride plastic. Other products that contain phthalates are vinyl flooring, adhesives, building materials, plastic bags, food packaging, garden hoses, inflatable recreational toys, blood-storage bags, and intravenous medical tubing. Animal studies show that fetal exposure to phthalates causes developmental toxicity in the male reproductive system. In baby boys who were exposed in the womb, researchers have seen an association between high levels of maternal phthalate metabolites—seen in about 25 percent of U.S. women—and effects on the measurement of an important health-related marker known as anogenital distance. Researchers have associated high levels of phthalates with lower sperm motility in adult men, and one study has correlated phthalates with abdominal obesity and insulin resistance in men.

Are you exposed?

Phthalates are widely detected in the U.S. population. Women are slightly more exposed than men, and younger

children between the ages of six and eleven are more exposed than older children. Exposure can occur through
direct use of products that contain phthalates and through
eating and breathing. Phthalates are metabolized and
excreted quickly from the body, but we experience new
exposures many times each day.

What to do?

Check the label of your personal care products and don't
buy those that use phthalates as an ingredient. Another
word to watch for on the ingredient label is "fragrance."
Avoid it if you see it listed in detergents, lotions, or soaps.
Why? Because diethyl phthalate (DEP) is an ingredient in
many fragrances, although it does not have to be listed
separately on the label. Don't use plastic containers in the
microwave because it can cause phthalates in the plastic to
leach into your food. Remember that phthalates are what
make vinyl soft. So stay away from PVC plastic in toys,
shower curtains, floor coverings, and building materials.
There are alternatives: Ethylene vinyl acetate, or EVA,
which does not require a plastic softener, is being used by
some companies as a replacement for soft PVC. Other
companies have switched to polyethylene or other plastic
polymers that do not require phthalates. Paints and other
hobby materials may contain phthalates as solvents, so
make sure you use them in a well-ventilated area. And last,
think twice about air fresheners. Many brands contain
phthalates, though the ingredient does not have to be
listed on the label.

CHEMICAL: POLYBROMINATED DIPHENYL ETHERS (PBDEs)

PBDEs are a family of flame retardants and a chemical cousin of the long-banned industrial substances known as PCBs. Beginning in the 1970s, PBDEs were added to up-holstered furniture, mattresses, carpet padding, vehicle up-holstery, and electronics. Two types of PBDEs, Octa and Penta, were withdrawn from the market in 2005 because of concerns about widespread exposures and toxicity. The most widely used member of this chemical family—the flame retardant known as Deca—remains on the market for use primarily in electronics. Scientists are concerned that Deca breaks down to form the types of PBDEs already taken off the market. In test animals, PBDEs cause an array of health effects, including thyroid hormone dis-ruption, permanent learning and memory impairment, be-havioral changes, hearing deficits, delayed puberty onset, decreased sperm count, and fetal malformations.

Are you exposed?

Fat-loving PBDEs are highly persistent in people and the environment. The chemicals build up in the body, are stored in fatty tissues and body fluids such as blood and breast milk, and can be passed on to fetuses and infants during pregnancy and lactation. People are primarily exposed to PBDEs in their homes, offices, and vehicles. Secondary sources are foods, primarily meat, dairy, fish, and eggs.

What to do?

Look for PBDE-free electronics and furniture. PBDEs should not be in mattresses, couches, and other foam products sold in 2005 or later. However, they are still put in some new televisions and computer monitors. If you're not sure if a product contains PBDEs, contact the manufacturer and ask. Avoid contact with crumbling old foam from carpet padding, old mattress pads, and stuffed furniture. If you can't replace the items, tightly tape over the tears or rips. Isolate an area when replacing old carpet padding so you don't spread dust. Vacuum regularly. For a list of PBDE-free products, see the Environmental Working Group's list at www.ewg.org/pbdefree.

CHEMICAL: BISPHENOL A

Bisphenol A is the chemical building block for polycarbonate plastic and epoxy resins. Dozens of animal studies suggest it disrupts the endocrine system at the low levels of exposure frequently measured in humans. Scientists note how recent trends in human disease relate to adverse effects observed in lab animals exposed to low levels of bisphenol A. Specific examples include prostate and breast cancer, urogenital abnormalities in baby boys, a decline in semen quality, early onset of puberty in girls, Type 2 diabetes, obesity, and neurobehavioral problems such as attention-deficit/hyperactivity disorder.

Are you exposed?

Canned foods lined with an epoxy resin are a common source of exposure to bisphenol A. So are food and beverage containers made from polycarbonate plastic. This includes baby bottles, sippy cups, water cooler jugs, and hard, clear, reusable polycarbonate drinking bottles many people use to cut down on waste. Some dental sealants and composites also contribute to low-level exposure.

What to do?

Limit the amount of canned food in your diet unless you are certain the producer is using bisphenol A–free containers. If you're not sure, call the company's consumer hotline. Look out for the number 7 stamp in the recycling symbol on plastics. Most—but not all—number 7 plastics are made from polycarbonate. But keep in mind that some alternatives to bisphenol A–derived containers, including Camelbak's and Nalgene's new lines of reusable bottles, also bear the number 7 stamp. Don't use heavily worn or scratched polycarbonate containers, as bisphenol A leaches more easily when the container starts to degrade. Think about replacing polycarbonate containers with aluminum, copolyester plastic, or stainless steel (though make sure the container is not lined with bisphenol A–containing resin). Buy bisphenol A–free baby bottles. Ask your dentist about the sealants and composites he or she uses.

CHEMICAL: PERFLUORINATED CHEMICALS

Perfluorinated chemicals resist grease, water, and stains. They've been used in consumer products since the 1950s and are associated with many famous names, including Teflon, Scotchgard, Stainmaster, and GORE-TEX. The most widely used and studied among the many different PFCs are the chemicals known as PFOS (perfluorooctane

sulfonate) and PFOA (perfluorooctanoic acid). The use of
these two chemicals has led to worldwide contamination
of people and places. Perfluorinated chemicals are extremely persistent. PFOA is used in manufacturing the
nonstick coating in Teflon. It's also a breakdown product
of stain- and grease-proof coatings. PFOA does not degrade. It has a half-life in the human body of nearly four
years. Other PFCs break down and turn into PFOA.
PFOS, which was in the Scotchgard formulation until
2000, has a half-life of more than five years. New epidemiological studies suggest that maternal exposure to perfluorinated chemicals may lower a baby's birth weight and
contribute to infertility.

Are you exposed?
Humans are exposed through contaminated water and
food, and by breathing contaminated air.

What to do?
When buying new cookware, avoid items with nonstick
coatings; look for stainless steel, hard anodized aluminum,
and cast iron as alternatives. Decline optional treatments
for stain and dirt resistance on clothing, shoes, and furniture. Ask retailers to help you find items that have not
been pretreated with chemicals. Avoid fast-food packaging
and microwave popcorn bags because many are coated
with perfluorinated chemicals to inhibit grease stains.

Some groups on this list are famous. Others are probably not familiar. All are worth a visit.

Breast Cancer Fund offers a legislative tool kit with many ideas for local organizations. The site also provides links to information about eliminating the environmental causes of breast cancer. www.breastcancerfund.org.

Campaign for Safe Cosmetics aims to protect the health of consumers and workers by requiring the health and beauty industries to phase out the use of chemicals linked to cancer, birth defects, and other health problems, and replace them with safer alternatives. www.safecosmetics.org.

Center for Environmental Health works to eliminate toxics, support communities, strengthen laws, and make industries green. This small nonprofit was onto lead in children's toys long before the rest of the world figured it out. www.cehca.org.

Center for International Environmental Law aims to solve environmental problems and promote sustainability through the use of international laws and institutions. www.ciel.org.

Children's Environmental Health Network is dedicated to protecting the fetus and children from environmental hazards. It provides resources for health-care professionals and consumers alike. www.cehn.org.

Clean Production Action helps businesses develop strategies that are good for human health and the environment. www.cleanproduction.org.

Collaborative on Health and the Environment is a diverse network of 2,900 individual and organizational partners in forty-five countries working collectively to advance knowledge and promote action addressing the links between human health and environmental factors. Part of its efforts include a biomonitoring resource center. www.healthandenvironment.org.

Environment California Research & Policy Center is a statewide advocacy organization that investigates environmental and public-health problems, offers solutions, and helps educate the public and decision makers. It was a leading proponent of California's 2007 law banning phthalates from toys and other products for children. www.environmentcalifornia.org.

Environmental Defense works directly with businesses, governments, and communities to find the best solutions to environmental problems. www.environmental defense.org.

Environmental Health News aggregates daily links to

articles in the international press about environmental health. A fabulous resource. www.environmentalhealth news.org.

Environmental Health Strategy Center focuses on protecting human health by reducing our exposure to toxic chemicals and promoting safer alternatives. www.preventharm.org.

Environmental Working Group was among the first advocacy organizations to call attention to the issue of human body burden by sponsoring biomonitoring studies detailing womb-to-tomb exposures. The EWG website contains a treasure trove of information about toxics commonly found in consumer products. www.ewg.org.

Greenpeace, the granddaddy of environmental watchdog groups, is concerned about pollution in people, too. www.greenpeace.org/usa/campaigns/toxics.

Health Care Without Harm is an international coalition of nearly five hundred organizations working to transform health care so that it poses no harm to humans or the environment. Because of the organization's efforts, some of the largest health-care providers in the world no longer use PVC plastics and brominated flame retardants. www.noharm.org.

Making Our Milk Safe (MOMS) was founded in 2005 by four nursing mothers with the mission of protecting the health of babies by eliminating the growing threat of toxic chemicals and industrial pollutants in human breast milk. MOMS is building a movement of mothers—and others—to speak out against the presence of toxics in our environment, our bodies, and our breast milk. www.safemilk.org.

National Environmental Trust runs education campaigns

on a variety of subjects, including environmental health. Its website contains a primer on why our federal toxics policy needs updating. www.net.org.

Natural Resources Defense Council, one of the nation's largest environmental advocacy groups, has an interactive website that can help you reduce your daily exposures. www.nrdc.org.

Northwest Coalition for Alternatives to Pesticides offers free information sheets on pesticide-free alternatives to pest problems. www.pesticide.org.

Pesticide Action Network North America works for socially just and sustainable agricultural practices. www.panna.org.

Physicians for Social Responsibility has made toxics and health one of its cornerstones. Its website offers information sheets and reports on the topic. www.psr.org.

Safer States is a network of organizations around the country that champions solutions to protect the public from toxic chemicals. www.saferstates.com.

Science and Environmental Health Network encourages the practice of science in the public interest and advocates for the precautionary principle. www.sehn.org.

Sightline Institute, formerly Northwest Environment Watch, is a regional sustainability think tank with a special interest in toxics. www.sightline.org.

Silent Spring Institute researches the links between the environment and women's health, particularly breast cancer. www.silentspring.org.

WWF Toxics promotes programs to control toxics globally. The website offers many consumer tips. www.worldwildlife.org/toxics.

Appendix 3:
LEARN MORE FROM
GOVERNMENT SOURCES

If you want to dig deeper into biomonitoring and how it is used, these resources will get you started.

The Centers for Disease Control and Prevention is tracking human exposure to common chemicals through the use of biomonitoring. It issues national reports approximately every two years. To order the latest report or for information about specific chemicals, contact:

CDC
Division of Laboratory Sciences
Mail Stop F-20
4770 Buford Highway, NE
Atlanta, GA 30341-3724
800-232-4636
cdcinfo@cdc.gov
www.cdc.gov/biomonitoring

If you want to learn more about chemicals of concern, the National Institute of Environmental Health Sciences is a good starting point. It conducts and coordinates research into how environmental exposures, like those from chemicals contained in everyday products, influence the development and progression of human disease.

NIEHS
PO Box 12233
Mail Drop B2-02
Research Triangle Park, NC 27709
919-541-3345
bruskec@niehs.nih.gov
www.niehs.nih.gov

Within NIEHS, the National Toxicology Program evaluates substances of public-health concern by developing and applying tools of modern toxicology and molecular biology. It is a leading resource for information about potentially hazardous effects of chemicals on human reproduction and development. For information about NTP study results and projects, go to www.ntp.niehs.nih.gov.

One noteworthy NTP program is the Center for Evaluation of Risks to Human Reproduction. For general questions regarding the CERHR or for personal questions regarding the possible effects of exposures on fertility, pregnancy, or unborn children, contact:

Dr. Michael D. Shelby
NIEHS EC-32
PO Box 12233
Research Triangle Park, NC 27709
919-541-3455
shelby@niehs.nih.gov

For links to CERHR reports, including its highly controversial evaluation of bisphenol A, go to www.cerhr .niehs.nih.gov.

Just under way is the National Children's Study, by far the most ambitious research yet undertaken to examine the effects of environmental influences on the health and development of children. It follows more than a hundred thousand U.S. children from birth until age twenty-one with the goal of improving the health and well-being of all children. For details see www.nationalchildrensstudy.gov.

 **Notes**

In some instances, where the text is clear about the source, I have noted where to access additional information. Addresses are current as of February 2009.

Introduction: Coming Clean

4 *In short, the United States does not have . . .*   For a brilliant analysis of this crisis, see Susanne Rust, Meg Kissinger, and Cary Spivak, "Are Your Products Safe? You Can't Tell," *Milwaukee Journal Sentinel*, November 24, 2007, at www.jsonline.com/watchdog/watchdogreports/29331224.html.

5 *Meanwhile, worried pet owners besieged . . .*   See *FDA Science and Mission at Risk: A Report of the Subcommittee on Science and Technology*, FDA Science Board, November 2007. www.fda.gov/ohrms/dockets/ac/07/briefing/2007-4329b_02_01_FDA%20Report%20on%20Science%20and%20Technology.pdf.

5 *The resulting reductions in the blood lead levels . . .*   For more background see the EPA's lead awareness program at www.epa.gov/lead.

6 *And Congress has responded . . .*   See: "The President and Product Safety," *The New York Times*, August 5, 2008, www.nytimes.com/2008/08/05/opinion/05tue2.html.

7 *Indeed, the federal toxics law* discourages . . . For more of Goldman's assessment, see the transcript of "Chemicals: Making Public Health Policy in the Face of Uncertainty," George Washington
University School of Public Health and Health Services, November 27, 2007, at www.kaisernetwork.org/health_cast/hcast_
index.cfm?display=detail&hc=2412.

9 *Acknowledging the breathtaking scope* . . . See the full text of
Von Eschenbach's remarks on February 29, 2008, at www.fda
.gov/oc/speeches/2008/fdaworld022908.html.

9 *A scathing 2007 report* . . . See *FDA Science and Mission at Risk.*

9 *As the FDA founders* . . . See Annys Shin, "Goodbye to Bob,"
Washington Post, January 4, 2008, www.washingtonpost.com/
wp-dyn/content/article/2008/01/04/AR2008010403778.html;
and Elizabeth Williamson, "Safety Chief Defends Record on
Toys," *Washington Post*, January 10, 2008, www.washingtonpost
.com/wp-dyn/content/article/2008/01/09/AR2008010903176
.html.

10 *Congress moved to* . . . See details of the Consumer Product
Safety Improvement Act at www.cpsc.gov/ABOUT/Cpsia/
legislation.html.

1. A Chemical Stew: Body Burden

13 *Intellectually, she understands as well as anyone* . . . I visited
Sharyle Patton at her home and office in Bolinas in late 2003, just
as I was beginning my research for this book. Her background information was collected in personal and telephone interviews
during 2003 and 2004. Patton, now the director of Commonweal's Biomonitoring Resource Center, was among nine volunteers in a biomonitoring study conducted by researchers from
Mount Sinai. Study results can be found at archive.ewg.org/
reports/bodyburden1.

13 *These toxic substances accumulate* . . . The CDC issues a *National Report on Human Exposure to Environmental Chemicals*
approximately every two years. For background about the report
and how it is used by researchers and government agencies, see
www.cdc.gov/exposurereport/report_factsheets.htm.

14 *That water-repellent jacket* . . . More information and an EPA
fact sheet are available at www.epa.gov/oppt/pfoa/pubs/pfoainfo
.htm.

14 *That cute yellow bath toy* . . . The *San Francisco Chronicle* tested
sixteen children's products for the presence of phthalates. See
Jane Kay, "San Francisco Prepares to Ban Certain Chemicals in

Products for Kids, but Enforcement Will Be Tough—and Toy-makers Question Necessity," *San Francisco Chronicle*, November 19, 2006, www.sfgate.com/cgi-bin/article.cgi?f=/c/a/2006/11/19/TOXICTOYS.TMP. San Francisco halted the ban after the chemical industry sued. However, on October 15, 2007, Governor Arnold Schwarzenegger signed a bill banning phthalates in children's products sold in California. See "California OKs Phthalates Ban on Children's Products," Reuters, www.reuters.com/article/healthNews/idUSN1443724320071015.

14 *That TV you spend hours . . .* See Florence Williams, "Toxic Breast Milk?" *New York Times Magazine*, January 9, 2005, www.nytimes.com/2005/01/09/magazine/09TOXIC.html. In April 2007, despite strong industry opposition, Washington became the first state to pass legislation banning the use of the most common PBDE flame retardant, Deca. See www.watoxics.org/pressroom/press releases/pbde-victory.

14 *"It's overwhelming" . . .* The Environmental Working Group has incorporated its biomonitoring research into a fascinating interactive Web tool called the Human Toxome Project that maps the links between chemicals and disease. See www.bodyburden.org.

15 *In the United States, our chemical neighborhood . . .* The Government Accountability Office issued reports in 2005, 2006, and 2007 that discuss the limitations of the Toxic Substances Control Act. They are, respectively, GAO-05-458, GAO-06-1032T, and GAO-07-825. The reports also contain suggestions for how to strengthen U.S. laws governing the use of toxic chemicals. See www.gao.gov.

16 *Who can argue . . .* For the latest ads, visit www.americanchemistry.com.

17 *Examples originate from all over . . .* The CDC's *Third National Report on Human Exposure to Environmental Chemicals* presents exposure data from 148 environmental chemicals. It is NCEH Pub. 05-0570. www.cdc.gov/exposurereport.

17 *My own body burden analysis . . .* Glenys Webster, who leads the Chemicals, Health and Pregnancy study launched in 2007 at the University of British Columbia, kindly interpreted my biomonitoring results for me. Her study, expected to be completed in late 2008, focuses on how flame retardants and perfluorinated compounds used as stain or water repellents affect maternal thyroid function, which in turn could affect fetal development. www.cher.ubc.ca/News/MediaReleases/chirp-11-01-07.asp.

19 *The EPA proposed a "significant new use rule" . . .* For a history

of the agency's recent actions regarding perfluorinated chemicals, see www.health.state.mn.us/divs/eh/hazardous/topics/pfcwork shop0507/pfcsepa.pdf. Minnesota, where 3M manufactured perfluorinated chemicals for four decades, is especially concerned about pollution from the substances. A state health department fact sheet offers answers to consumer-related questions, at www .health.state.mn.us/divs/eh/hazardous/topics/pfcshealth.hml.

20 *Certainly, these substances are as stubborn* . . . The United Nations Environment Programme has published a very readable guide to the Stockholm Convention, which aims to rid the world of twelve persistent organic pollutants. www.pops.int/ documents/guidance/beg_guide.pdf. The United States has yet to adopt the Stockholm Convention; see Kristin S. Schafer, "One More Failed U.S. Environmental Policy," at www.fpif.org/ fpiftxt/3492.

21 *"Biomonitoring is not just"* . . . I visited the CDC's biomonitoring laboratory in Atlanta in January 2004. My interviews with Jim Pirkle were conducted in person and by telephone in January and March 2004.

25 *As the ability to pinpoint pollutants* . . . Telephone interview with Dr. Richard Jackson, March 2004.

25 *For years now, it's been well understood* . . . Genes alone do not tell the whole story. Recent increases in chronic diseases like diabetes, childhood asthma, obesity, and autism are associated with changes in our environments, diets, and activity levels, which may produce disease in genetically predisposed persons. Information about the NIH's Genes, Environment and Health Initiative can be found at www.gei.nih.gov/index.asp. Also see the website maintained by Duke University researcher Randy Jirtle, www .geneimprint.com.

26 *Since the beginning of the chemical revolution* . . . For background on this paradigm shift, see the report by Pete Myers and Wendy Hessler at www.environmentalhealthnews.org/science background/2007/2007-0415nmdrc.html.

27 *The state of the research was summarized* . . . The report is available at www.niehs.nih.gov/health/topics/agents/endocrine/docs/ endocrine.pdf.

28 *No one is exposed to just one* . . . Michael P. Wilson et al., *Green Chemistry in California: A Framework for Leadership in Chemicals Policy and Innovation,* California Policy Research Center, 2006, www.ucop.edu/cprc/documents/greenchemistryrpt.pdf.

28 *Indeed, two of the world's leading researchers* . . . P. Grandjean and P. J. Landrigan, "Developmental Neurotoxicity of Industrial Chemicals—A Silent Pandemic," *Lancet* 368, issue 9353 (December 2006): 2167–78.

29 *"The bad news is"* . . . My interviews with Myers, a seminal figure in the environmental health movement since he coauthored *Our Stolen Future* (New York: Dutton, 1996), took place in person and by telephone between 2003 and 2007. This quote is from his lecture at the Seattle Art Museum on January 27, 2004, "A New View on Toxic Chemicals and How They Impact Our Health."

2. Chemicals We've Loved: Consumer Conveniences

33 *By the mid-twentieth century* . . . To glimpse the wondrous way things were at Disneyland when the park opened on July 17, 1955, see www.yesterland.com/dl1955.html. By virtually every economic measure, enthused *Time* magazine, 1956 was the greatest year in history. Its ebullient review is at www.time.com/time/magazine/article/0,9171,867536,00.html.

36 *Most certainly, scores of synthetics* . . . Several books informed my reporting about the proliferation of chemicals during the mid-twentieth century. For details about production, I relied on Conrad Berenson, *The Chemical Industry: Viewpoints & Perspectives* (New York: Interscience Publishers, 1963); Sheldon Hochheiser, *Rohm and Haas: History of a Chemical Company* (Philadelphia: University of Pennsylvania Press, 1986); and Peter H. Spitz, *Petrochemicals: The Rise of an Industry* (New York: John Wiley, 1988). For an understanding of the social and cultural implications of plastic, which transformed the nature of American life, I found Stephen Fenichell, *Plastic: The Making of a Synthetic Century* (New York: Harper, 1996); and Jeffrey L. Meikle, *American Plastic: A Cultural History* (New Brunswick, NJ: Rutgers University Press, 1995) terrifically helpful. For a critical appraisal of the chemical industry, I read Cathy Trost, *Elements of Risk: The Chemical Industry and Its Threat to America* (New York: Times Books, 1984). For industry data, the American Chemistry Council's *Guide to the Business of Chemistry* was without peer.

39 *The roots of this modern material world* . . . A visit to the Capitol Mall and a Smithsonian Institution exhibit, "Science in American Life," helped me understand how chemistry has fused with moments in history, creating cultural and political turning points.

40 *Today, the chemical industry is the largest single user* . . . As noted by researchers at Lawrence Berkeley National Laboratory, the chemical industry does not trumpet its voracious energy appetite; there is not much information about the chemical industry's energy use and energy intensity in the public domain. See industrial-energy.lbl.gov/node/86.

40 *Inside a petrochemical plant* . . . For an overview of industry processes and challenges, see *OECD Environmental Outlook for the Chemicals Industry*, prepared by the Organisation for Economic Co-operation and Development, www.oecd.org/dataoecd/7/45/2375538.pdf. Also see *Green Chemistry in California: A Framework for Leadership in Chemicals Policy and Innovation*, www.ucop.edu/cprc/documents/greenchemistryrpt.pdf.

41 *These breakthroughs were based on organochlorines* . . . For a detailed and thoughtful accounting of why organochlorines are a global health hazard, I relied on Joe Thornton, *Pandora's Poison: Chlorine, Health, and a New Environmental Strategy* (Cambridge, MA: MIT Press, 2001).

43 *All of the "dirty dozen" substances* . . . See www.pops.int/documents/guidance/beg_guide.pdf.

44 *Chemicals underlie the consumer culture* . . . See www.ucop.edu/cprc/documents/greenchemistryrpt.pdf, p. 11.

46 *In effect, the toxics legislation* . . . Lynn R. Goldman, "Preventing Pollution? U.S. Toxic Chemicals and Pesticides Policies and Sustainable Development," *Stumbling Toward Sustainability*, ed. John C. Dernbach (Washington, DC: Environmental Law Institute, 2002), chap. 17.

49 *In comparison to the fate of asbestos* . . . The Environmental Working Group's report *The Asbestos Epidemic in America*, reports.ewg.org/reports/asbestos/facts.

51 *"We need to shift the responsibility"* . . . Wallström, then European commissioner for the environment, in a keynote speech at the United States–European Union Transatlantic Environment Conference on Chemicals, Charlottesville, Virginia, April 26, 2004.

52 *It even enlisted the Bush administration's help* . . . Cheryl Hogue, "Waxman Criticizes Lobby Efforts Abroad," *Chemical & Engineering News*, April 6, 2004, pubs.acs.org/cen/news/8214/8214reach.html.

52 *Greg Lebedev* . . . "Chemistry Means Business," keynote address for Pittsburgh Chemical Day, May 11, 2004.

53 *At a national conference on tetraethyl lead* . . . Sheldon Rampton and John Stauber, *Trust Us, We're Experts! How Industry Manip-*

ulates Science and Gambles with Your Future (New York: Jeremy Tarcher, 2001), 93.

53 *At Chicago's Century of Progress* . . . Roland Marchand and Michael L. Smith, "Corporate Science on Display," *Scientific Authority and Twentieth-Century America*, ed. Ronald G. Walters (Baltimore: Johns Hopkins University Press, 1997), 160–62, 164–65.

55 *Those who questioned the wisdom* . . . For a meticulous, readable biography of Rachel Carson, I recommend Linda Lear, *Rachel Carson: Witness for Nature* (New York: Henry Holt, 1997). More than three decades after the publication of *Silent Spring*, Carson remains a popular target for criticism from the right wing. See writings by Herbert I. London, Dennis T. Avery, and Michael Fumento at the Hudson Institute, www.hudson.org. Also see writings by Elizabeth M. Whelan at the American Council on Science and Health, www.acsh.org.

3. Kermit's Blues: Atrazine and Frogs

57 *In 1998, when Tyrone Hayes* . . . My interviews with Tyrone Hayes took place between 2004 and 2008. Because of his controversial research, Hayes has been the subject of several lengthy profiles, including Goldie Blumenstyk, "The Story of Syngenta & Tyrone Hayes at UC Berkeley: The Price of Research," *Chronicle of Higher Education* 50, issue 10 (2003), www.mindfully.org/Pesticide/2003/Syngenta-Tyrone-Hayes310octo3.htm; William Souder, "It's Not Easy Being Green: Are Weedkillers Turning Frogs Into Hermaphrodites?" *Harper's*, August 23, 2006, www.biologicaldiversity.org/swcbd/programs/science/pesticides/bay-area.html; Elizabeth Royte, "Transsexual Frogs," *Discover*, February 2003, discovermagazine.com/2003/feb/featfrogs. All quotations from Hayes were derived from my interviews with him, his testimony before state or federal agencies, or his presentations at scientific conferences.

57 *A decade or so earlier* . . . See the EPA's consumer fact sheet on atrazine, www.epa.gov/safewater/dwh/c-soc/atrazine.html.

60 *Hormones, released by endocrine glands* . . . To learn more about the endocrine system and the interplay between hormones and the environment, visit e.hormone hosted by the Center for Bioenvironmental Research at Tulane and Xavier universities, e.hormone.tulane.edu.

60 *Environment Canada* . . . See the agency's web page on endocrine disruptors, www.ec.gc.ca/eds/fact/broch_e.htm.

62 *But Dana Barr* . . . Personal and telephone interviews with Dana Barr conducted in 2004.

66 *The EU restrictions were instituted* . . . See the final European Commission review report, ec.europa.eu/food/plant/protection/ evaluation/existactive/list_atrazine.pdf. For a transcript of a "Living on Earth" radio broadcast featuring Hayes discussing the EU ban, go to www.loe.org/shows/segments.htm?programID=06-P13-00016&segmentID=1.

66 *About three-fourths of U.S. acreage* . . . See the atrazine website created by the industry-allied Center for Regulatory Effectiveness, www.thecre.com/atrazine/use.htm.

67 *As Syngenta toxicologist* . . . See Juliet Eilperin, "High Weedkiller Levels Found in River Checks," *Washington Post*, December 9, 2007, page A6.

67 *Chemists at J. R. Geigy* . . . For background on atrazine's corporate lineage, I relied on Dan Fagin and Marianne Lavelle, *Toxic Deception: How the Chemical Industry Manipulates Science, Bends the Law and Endangers Your Health* (Monroe, ME: Common Courage Press, 1999).

67 *A 1991 USGS study* . . . See the EPA's archived notes on atrazine, www.epa.gov/OWOW/info/NewsNotes/pdf/25issue.pdf.

68 *A 2006 USGS report* . . . *Pesticides in the Nation's Streams and Ground Water, 1992–2001*, pubs.usgs.gov/fs/2006/3028.

68 *Moreover, monitoring by Syngenta* . . . See Eilperin, "High Weedkiller Levels Found in River Checks."

69 *In 2000, studies supported by Syngenta* . . . For the background of the EPA's decision, see www.epa.gov/scipoly/sap/meetings/2000/062700_mtg.htm.

69 *This is the point at which the relationship* . . . Hayes outlined his complaints in his resignation letter to the Ecorisk panel, which he shared with me and during our interviews. Dr. Tim Pastoor, a toxicologist and head of Syngenta's global risk assessments, denied Hayes's complaints in my 2005 telephone interviews with him.

70 *In 2002, just months before the EPA* . . . Tyrone Hayes et al., "Herbicides: Feminization of Male Frogs in the Wild," *Nature* 419, no. 9910 (2002): 895–96, and Tyrone Hayes et al., "Hermaphroditic, Demasculinized Frogs after Exposure to the Herbicide Atrazine at Low Ecologically Relevant Doses," *Proceedings of the National Academy of Sciences* 99, no. 8 (2002): 5476–80.

70 *Syngenta and the expert panel* . . . Syngenta has removed the press release from its website. It is accessible through the Kansas State University food safety archive, archives.foodsafety.ksu.edu/

agnet/2002/6-2002/agnet_june_20-2.htm#FROG%20RESEARCH
%20ON.

71 *A report Hayes published . . .* Tyrone Hayes, "There Is No Denying This: Defusing the Confusion About Atrazine," *BioScience* 54,
no. 12 (December 2004): 1138–49, www.iceh.org/pdfs/SBLF/
HayesBioscience.pdf.

72 *Pete Myers . . .* Myers is the former executive director of the
W. Alton Jones Foundation, which helped fund Hayes's research
after he severed his ties with the Syngenta-funded panel.

72 *For years, big tobacco disputed . . .* Peter Montague, "Ending Government Regulation by Manufacturing Doubt," *Rachel's
Environment & Health News*, 824 and 825, www.rachel.org/
en/newsletters/search/rachels_news.

72 *Syngenta's home page on atrazine . . .* See Steven Milloy, "Frog
Study Leaps to Conclusions," Fox News, April 19, 2002, www
.syngentacropprotection-us.com/prod/herbicide/Atrazine. The
site linked Milloy's commentary as of May 6, 2006; it has since
been removed and replaced by a cover story from *AgriMarketing*
called "Atrazine: Legendary Marketing of a Legendary Molecule."

73 *Tom Steeger, a senior biologist . . .* Steeger's final report is *White
Paper on Potential Developmental Effects of Atrazine on Amphibians.* The paper and other background information on EPA's regulatory actions and decisions for atrazine are available at the
atrazine docket, EPA-HQ-OPP-2007-0498, www.regulations.gov.

77 *Filed in late 2002 . . .* For further background on Jim Tozzi and
the Center for Regulatory Effectiveness, see Chris Mooney,
The Republican War on Science (New York: Basic Books: 2005),
102–20.

77 *"If accepted," wrote Jennifer Sass . . .* See the entire letter, and a
response from the Center for Regulatory Effectiveness, *Environmental Health Perspectives* 112, 1 (2004), at www.ehponline.org/
docs/2004/112-1/correspondence.html.

78 *Chief among them . . .* Rick Weiss, " 'Data Quality' Law Is
Nemesis of Regulation," *Washington Post*, August 16, 2004,
page A1.

79 *Based on data submitted by Syngenta . . .* See the EPA's updates
on activities related to atrazine's reregistration at www.epa.gov/
oppsrrd1reregistration/atrazine/atrazine_update.htm.

82 *Today, undergraduates . . .* Dinan's case teaching notes are available at www.sciencecases.org/kermit/kermit.pdf.

84 *In November 2003, Hayes rocked . . .* Rebecca Renner, "Contro-

versy Clouds Atrazine Studies," *Environmental Science & Technology* 38, no. 6 (2004): 107A–108A.

84 *In the end . . .* See Statement of Anne E. Lindsay, Deputy Director, Office of Pesticide Programs U.S. Environmental Protection Agency, before the Agriculture and Rural Development Committee of the Minnesota House of Representatives, February 16, 2005, www.house.leg.state.mn.us/hinfo/swkly/2005-06/sw705.pdf.

85 *He has continued his research . . .* See Tyrone Hayes et al., "Characterization of Atrazine-Induced Gonadal Malformations in African Clawed Frogs," *Environmental Health Perspectives* 114, suppl. 1 (2006): 134–41, www.ehponline.org/members/2006/8067/8067.html, and Tyrone Hayes et al., "Pesticide Mixtures, Endocrine Disruption, and Amphibian Declines: Are We Underestimating the Impact?" *Environmental Health Perspectives* 114, suppl. 1 (2006): 40–50, www.ehponline.org/members/2006/8051/8051.html.

86 *The U.S. Geological Survey shares . . .* See Gilliom's remarks in a news release, water.usgs.gov/nawqa/pnsp.

86 *Atrazine in particular . . .* Ibid., 16.

4. What Price Beauty? Phthalates and You

88 *On May 17, 1933 . . .* For the history of cosmetics, including the story of how the blinding effects of Lash Lure contributed to the Food, Drug, and Cosmetic Act, I relied on Teresa Riordan, *Inventing Beauty: A History of the Innovations That Have Made Us Beautiful!* (New York: Broadway Books, 2004), Gwen Kay, *Dying to Be Beautiful: The Fight for Safe Cosmetics* (Columbus: Ohio State University Press, 2005), and Kathy Peiss, *Hope in a Jar: The Making of America's Beauty Culture* (New York: Metropolitan Books, 1998).

90 *The average adult . . .* Estimates of the number of personal-care products used daily vary from source to source. I used the number cited by Morris Shriftman of Avalon Cosmetics.

90 *In 2004, the European Union . . .* See the EU's website on the cosmetics directive, ec.europa.eu/enterprise/cosmetics/index_en.htm.

91 *Skin sensitivity is the number one . . .* Dori Stehlin, "Cosmetic Safety: More Complex Than at First Blush," *FDA Consumer*, November 1991, updated May 1995, www.cfsan.fda.gov/~dms/cossafe.html.

91 *The cosmetics trade and lobby . . .* Visit the Personal Care Products Council website, www.personalcarecouncil.org.

92 *As the FDA's own website notes . . .* See the agency's page on its authority, www.cfsan.fda.gov/~dms/cos-pol.html.

92 *Before a chemist named Dr. John E. Bailey . . .* Judith E. Foulke, "Cosmetics Ingredients: Understanding the Puffery," *FDA Consumer*, May 1992, updated February 1995, www.fda.gov/fdac/reprints/puffery.html.

93 *The Environmental Working Group . . .* Skin Deep, the Environmental Working Group's interactive cosmetics ingredients database, is at www.ewg.org. For a copy of a letter EWG sent to the FDA notifying the agency of its findings, see www.ewg.org/node/22610.

95 *The campaign began . . .* For a detailed history see Stacy Malkan, *Not Just a Pretty Face: The Ugly Side of the Beauty Industry* (Gabriola Island, BC, Canada: New Society Publishers, 2007). See the Campaign for Safe Cosmetics at www.safecosmetics.org.

97 *Reformulating is expensive . . .* See Rosemary Carstens, "The Dark Side of Beauty," *Alternative Medicine*, January 2006, www.carstencommunications.com/articles/06DarkSideofBeauty.pdf.

101 *After the Campaign for Safe Cosmetics reported . . .* See the FDA's "Lipstick and Lead: Questions and Answers," December 27, 2007, www.cfsan.fda.gov/~dms/cos-pb.html, and the announcement of the senators' request for an investigation at www.kerry.senate.gov/cfm/record.cfm?id-287801.

103 *As the toxicologist . . .* Marc Lappé, *The Body's Edge: Our Cultural Obsession with Skin* (New York: Henry Holt, 1996).

104 *We breathe them . . .* See the CDC's "Spotlight on Phthalates," www.cdc.gov/exposurereport/pdf/factsheet_phthalates.pdf.

104 *Using phthalates restrictions enacted . . .* See Feinstein's announcement of the ban at www.feinstein.senate.gov/public/index.cfm?FuseAction=NewsRooms.PressReleases&ContentRecord_id-6F0815F-D7CF-C498-3DEC-85DF3900CDAC.

106 *Toxicologists know from rat studies . . .* J. A. Hoppin, "Male Reproductive Effects of Phthalates: An Emerging Picture," *Epidemiology* 14 (May 2003): 259–60. Summary available at www.ourstolenfuture.org/NEWSCIENCE/reproduction/sperm/2003/2003-0519hoppin.htm.

106 *At the same time, rat . . .* According to a 2008 National Academy of Sciences report, the EPA underestimates the health risks of phthalates by analyzing each chemical one at a time. See: Marla Cone, "Scientists to EPA: Risks of chemicals that alter male hormones should be analyzed together," *Environmental Health News*, December 18, 2008, www.environmentalhealthnews.org/

ehs/news/scientists-to-epa-risks-of-chemicals-that-alter-male-
hormones-should-be-analyzed-together-to-protect-human-health-
national-panel-says.

107 *Scientists postulate* . . . N. E. Skakkeback et al., "Testicular
Dysgenesis Syndrome: An Increasingly Common Developmental Disorder with Environmental Aspects," *Human Reproduction*
16, no. 5 (May 2001): 972–78, humrep.oxfordjournals.org/cgi/
content/short/16/5/972.

107 *In the case of phthalates* . . . G. Latini et al., "In Utero Exposure
to di-(2-ethylhexyl)phthalate and Duration of Human Pregnancy," *Environmental Health Perspectives* 111, no. 14 (November 2003): 1783–85, www.pubmedcentral.nih.gov/article
render.fcgi?artid=1241724; R. Hauser et al., "Altered Semen
Quality in Relation to Urinary Concentrations of Phthalate
Monoester and Oxidative Metabolites," *Epidemiology* 17 (November 2006): 682–91, summary at www.ourstolenfuture.org/
newscience/oncompounds/phthalates/2006/2006-1101hauseretal
.html; S. Duty et al., "Phthalate Exposure and Human Semen
Parameters," *Epidemiology* 14 (May 2003): 269–77, summary at
www.ourstolenfuture.org/newscience/reproduction/sperm/2003/
2003-0519dutyetal.htm; R. Stalhut et al., "Concentrations of Urinary Phthalate Metabolites Are Associated with Increased Waist
Circumference and Insulin Resistance in Adult U.S. Males," *Environmental Health Perspectives* 115, no. 6 (June 2007), 876–82,
www.ehponline.org/docs/2007/9882/abstract.html.

108 *But in the spring of 2005* . . . S. Swan et al., "Decrease in
Anogenital Distance Among Male Infants with Prenatal Phthalate
Exposure," *Environmental Health Perspectives* 113, no. 8 (Au-
gust 2005): 1056–61, www.shswan.com.

109 *A follow-up study by researchers* . . . K. Marsee et al., "Estimated Daily Phthalate Exposures in a Population of Male Infants
Exhibiting Reduced Anogenital Distance," *Environmental Health
Perspectives* 114, no. 6 (June 2006): 805–809, www.ehponline
.org/docs/2006/8663/abstract.html.

110 *Olivia James* . . . James detailed her son's hypospadias in telephone interviews with me during 2004.

5. Up in Flames: Polybrominated Diphenyl Ethers

112 *In the 1970s* . . . In January 2004, Washington governor Gary
Locke directed the state to examine the use of PBDEs and recommend options for reducing their use. Part of the process included the January 2006 publication of the detailed report,

Washington State Polybrominated Diphenyl Ether (PBDE) Chemical Action Plan: Final Plan. I relied extensively on this report for background about the use of PBDEs and their growing threat to the environment and humans. It is available at www.ecy.wa.gov/biblio/0507048.html. I also found helpful *Decabromodiphenylether: An Investigation of Non-Halogen Substitutes in Electronic Enclosures and Textile Applications,* Lowell Center for Sustainable Development, April 15, 2005, sustainableproduction.org/publ.shtml#SustainableProduction. See also Elizabeth Grossman, *High Tech Trash: Digital Devices, Hidden Toxics, and Human Health* (Washington, DC: Island Press, 2006).

113 *According to the Bromine Science and Environmental Forum* . . . See the industry association's website, www.bsef.com.

114 *Levels of PBDEs in human fat* . . . See Washington state report, 11.

114 *While most research* . . . Kellyn S. Betts, "A New Record for PBDEs in People," *Environmental Science & Technology* 39, no. 14 (2005): 296A–298A.

115 *In 2004, a study* . . . "Flame Retardant Linked to House Dust," *ScienceDaily,* January 7, 2005, www.sciencedaily.com/releases/2005/01/050106110114.htm.

115 *In 2005, researchers* . . . "Household Dust Is Main Source of Flame Retardants in Humans," *ScienceDaily,* July 7, 2005, www.sciencedaily.com/releases/2005/07/050707062329.htm.

116 *A 2006 case study* . . . Douglas Fischer et al., "Children Show Highest Levels of Polybrominated Diphenyl Ethers in a California Family of Four: A Case Study," *Environmental Health Perspectives* 114, no. 10 (October 2006): 1581–84, www.ehponline.org/members/2006/8554/8554.html.

116 *But research published in 2007* . . . Janice Dye et al., "Elevated PBDE Levels in Pet Cats: A Sentinel for Humans?" *Environmental Science & Technology* 41, no. 18 (2007), 6350–56. See also Carol Potera, "Chemical Exposures: Cats as Sentinel Species," *Environmental Health Perspectives* 115, no. 12 (December 2007): A580, www.pubmedcentral.nih.gov/articlerender.fegi?artid=2137107.

118 *As with hundreds* . . . See *Polybrominated Diphenyl Ethers (PBDEs) Project Plan,* EPA, March 2006, www.epa.gov/oppt/pbde.

118 *If PBDE levels continue* . . . *Growing Threats: Toxic Flame Retardants & Children's Health,* Environment California, April 2003, www.environmentcalifornia.org/reports/environmental-health. See

also *Mother's Milk: Record Levels of Toxic Fire Retardants Found in American Mothers' Breast Milk*, Environmental Working Group, www.ewg.org/node/8416.

119 *Whispers about problems . . .* See Washington state report, 12. See also Muhammed Akmal Siddiqi et al., "Polybrominated Diphenyl Ethers (PBDEs): New Pollutants—Old Diseases," *Clinical Medicine & Research* 1, no. 4 (October 2003): 281–90, www.pubmedcentral.nih.gov/articlerender.fcgi?artid=1069057.

120 *Canadian experts discovered . . .* J. J. Ryan et al., "Recent Trends in Levels of Brominated Diphenyl Ethers in Human Milks from Canada," *Organohalogen Compounds* 58 (2002): 173–76.

120 *In the United States . . .* See Arnold Schechter et al., "Polybrominated Diphenyl Ethers (PBDEs) in U.S. Mothers' Milk," *Environmental Health Perspectives* 111, no. 14 (November 2003): 1723–29, www.ehponline.org/docs/2003/6466/abstract.html.

120 *That same year . . .* M. Petreas et al., "High Body Burdens of 2,2',4,4'-Tetrabromodiphenyl Ether (BDE-47) in California Women," *Environmental Health Perspectives* 111, no. 9 (July 2003): 263–72, www.ehponline.org/docs/2003/6220/6220.html.

120 *And the exposure news . . .* See Linda S. Birnbaum et al., "Polybrominated Diphenyl Ethers: A Case Study for Using Biomonitoring to Address Risk Assessment Questions," *Environmental Health Perspectives* 114, no. 11 (November 2006): 1770–75, www.ehponline.org/members/2006/9061/9061.html.

120 *And while it's true . . .* W. J. Rogan et al., "Should the Presence of Carcinogens in Breast Milk Discourage Breast Feeding?" *Regulatory Toxicology and Pharmacology* 13 (1991): 228–40.

121 *Authorities ranging from the American Academy . . .* See the CDC's breast-feeding website, www.cdc.gov/breastfeeding.

121 *Just after her oldest . . .* J. She et al., "Polybrominated Diphenyl Ethers (PBDEs) and Polychlorinated Biphenyls (PCBs) in Breast Milk from the Pacific Northwest," *Chemosphere* 67, no. 9 (April 2007): S307–17. Also see *Flame Retardants in the Bodies of Pacific Northwest Residents*, Sightline Institute, www.sightlineinstitute.org.

123 *For nursing mothers . . .* Gregory Dicum, "Mother Knows Best," *Grist*, November 6, 2006, www.grist.org/news/maindish/2006/11/06/dicum.

124 *In terms of toxicology . . .* Telephone interview with Arnold Schecter, November 2006.

125 *Together, Penta and Octa . . .* Unfortunately, research is raising

safety concerns about Chemtura's alternative to Penta, marketed as Firemaster 550. Heather M. Stapleton et al., "Alternate and New Brominated Flame Retardants Detected in U.S. Dust," *Environmental Science & Technology* 42, no. 18 (September 15, 2008): 6910–16.

128 *And then there's the issue of Deca* . . . Heather Stapleton, *Brominated Flame Retardants: Assessing DecaBDE Debromination in the Environment*, EPHA Environment Network, May 2006.

129 *In the summer of 2007* . . . See Marla Cone, "Outspoken Scientist Dismissed from Panel on Chemical Safety," *Los Angeles Times*, February 29, 2008, www.latimes.com/news/science/environment/la-me-epa29feb29,1,2980474.story.

130 *Stapleton, a young scholar* . . . See Kellyn S. Betts, "Unwelcome Guest: PBDEs in Indoor Dust," *Environmental Health Perspectives* 116, no. 5 (May 2008): A202–A208, www.pubmedcentral.nih.gov/articlerender.fegi?artid-2367657.

131 *The trouble is* . . . Telephone interview with Heather Stapleton, November 2006.

131 *BSEF, for example* . . . See the trade group's website, www.bsef.com.

137 *"I think that is overblown"* . . . "Toxic Trade-Off: Flame Retardant Risk Stokes Debate," *Fort Worth Star-Telegram*, December 4, 2006.

137 *Companies such as Dell* . . . See "Smart Shoppers' PBDE Card," *The Green Guide* 106 (January/February 2005), www.thegreenguide.com. Also see Clean Production Action, www.cleanproduction.org.

138 *Then, on April 1, 2008* . . . "Europe Restores Toxic Flame Retardant Ban Similar to Washington's," Washington State Department of Ecology press release, April 4, 2008, www.ecy.wa.gov/news/2008news/2008-080.html.

139 *The pressure on companies* . . . See Richard A. Liroff, "Benchmarking Corporate Management of Safer Chemicals in Consumer Products—A Tool for Investors and Senior Executives," *Corporate Environmental Strategy* 12, issue 1 (January/February 2005).

6. The Goods on Bad Plastic: Bisphenol A

141 *Dr. Fred vom Saal* . . . My interviews with vom Saal took place by telephone and in person between 2005 and 2007.

142 *In an unprecedented consensus* . . . F. vom Saal et al., "Chapel Hill Bisphenol A Expert Panel Consensus Statement: Integration

of Mechanisms, Effects in Animals and Potential to Impact Human Health at Current Levels of Exposure," *Reproductive Toxicology* 24, issue 2 (2007): 131–38.

143 *This, combined with . . .* Word leaked about possible regulatory action a few days before Canadian health minister Tony Clement announced on April 18, 2008, that the government wanted bisphenol A out of baby bottles and the linings of infant formula cans. As a result, a wave of Canada's biggest retailers, including Wal-Mart and Hudson's Bay Co., swept their shelves of most polycarbonate plastic products even before Clement spoke. See "Sears Canada, Rexall Pharmacies, London Drugs and Home Depot Canada Remove Bisphenol A Products from Shelves," Environmental Defence Press release, April 17, 2008, environmental defence.ca. Manufacturers moved quickly, too. Nalgene Outdoor Products, which found a new niche for unbreakable polycarbonate containers it originally made for laboratory use by marketing them to hikers, fitness enthusiasts, and students, said it was phasing out polycarbonate plastic bottles and would instead rely on a different type of plastic. See nalgene-outdoor.com. Just as quickly, a California lawyer filed a class action lawsuit against Nalgene. See "Nalgene Sports Bottle Maker Sued over Toxics Claims," Reuters, April 23, 2008, www.reuters.com/article/health News/idUSN2335756720080424.

143 *Take a quick inventory . . .* See the American Chemistry Council–sponsored website, www.bisphenol-a.org.

144 *So useful are polycarbonate . . .* See Bisphenol A Expert Panel reports by the Center for the Evaluation of Risks to Human Reproduction, cerhr.niehs.nih.gov/chemicals/bisphenol/ bisphenol-eval.html.

144 *It's not surprising . . .* See Endocrine Disruptors Group site maintained by vom Saal, endocrinedisruptors.missouri.edu/vom saal/vomsaal.html.

145 *In the laboratory . . .* Steven G. Gilbert, *A Small Dose of Toxicology: The Health Effects of Common Chemicals* (Boca Raton, FL: CRC Press, 2004).

147 *In a 2007 study spearheaded . . .* See *Bisphenol A: Toxic Plastic Chemical in Canned Food*, Environmental Working Group, March 5, 2007, www.ewg.org/reports/bisphenola.

147 *University of Cincinnati . . .* H. Le et al., "Bisphenol A is released from polycarbonate drinking bottles and mimics the neurotoxic actions of estrogen in developing cerebellar neurons,"

Toxicology Letters 176, no. 2 (208): 149–56. See www.science daily.com/releases/2008/01/080130092108.htm.

147 *Another study* . . . C. Brede et al., "Increased Migration Levels of Bisphenol A from Polycarbonate Baby Bottles After Dishwashing, Boiling and Brushing," *Food Additives and Contaminants* 20, no. 7 (2003): 684–89.

147 *One study estimated* . . . See Laura N. Vandenberg et al., "Human Exposure to Bisphenol A (BPA)," *Reproductive Toxicology* 24, no. 2 (November–December 2007): 139–77. Also *Toxic Baby Bottles: Scientific Study Finds Leaching Chemicals in Clear Plastic Baby Bottles*, Environment California, February 27, 2007, www.environmentcalifornia.org/reports/environmental-health/ environmental-health-reports/toxic-baby-bottles.

147 *And in a 2007* . . . As of January 2009, all five leading baby bottle brands offered bisphenol A–free products.

148 *A chemical analysis of paper* . . . Vandenberg, "Human Exposure to Bisphenol A."

149 *Indeed, the leaders* . . . "Committee Probe Finds FDA Used Industry Studies to Approve Chemical in Infant Formula Liners," Committee on Energy and Commerce press release, April 8, 2008.

149 *The polymer scientists* . . . Douglas Fischer, "The Bisphenol A Dilemma: How Is Government Responding? It's Not," *Oakland Tribune*, April 22, 2007.

150 *As a medication* . . . For more information, see the CDC's DES page, www.cdc.gov/DES.

150 *Retha Newbold* . . . R. R. Newbold et al., "Long-Term Adverse Effects of Neonatal Exposure to Bisphenol on the Murine Female Reproductive Tract," *Reproductive Toxicology* 24, no. 2 (2007): 253–58.

151 *Dr. Pat Hunt* . . . M. Susiarjo et al., "Bisphenol A Exposure In Utero Disrupts Early Oogenesis in the Mouse," *PLoS Genetics*, January 12, 2007. See also Elizabeth Grossman, "Two Words: Bad Plastic," *Salon*, August 2, 2007, www.salon.com.

151 *This was Hunt's second* . . . P. A. Hunt et al., "Bisphenol A Exposure Causes Meiotic Aneuploidy in the Female Mouse," *Current Biology* 13 (2003): 546–53.

153 *His research team discovered* . . . Liza Gross, "The Toxic Origins of Disease," *PLoS Biology*, June 26, 2007, www.plosjournals.org.

154 *Relying on scientific reports* . . . Ibid.

154 *In a sobering* . . . "Common Organic Compound Found in

Many Household Products May Pose Health Risk to Breast Cells,"
Science Daily, April 3, 2008, www.sciencedaily.com/releases/2008/
04/080401231554.htm.

155 *"The list of diseases"* . . . Pete Myers, "Good Genes Gone Bad,"
American Prospect, March 19, 2006, www.prospect.org/cs/
articles?=good_genes_gone_bad.

157 *In 1998, the National Toxicology Program* . . . See cerhr.niehs
.nih.gov.

158 *However, the panel's work* . . . Marla Cone, "Agency Linked to
Chemical Industry," *Los Angeles Times*, March 4, 2007.

158 *But the program let* . . . See the bisphenol A documents, including the *Expert Panel Draft Report by the Center for the Evaluation
of Risks to Human Reproduction* and public comments, at
cerhr.niehs.nih.gov/chemicals/bisphenol/bisphenol-eval.html.

161 *Sales of BornFree baby bottles* . . . See Elizabeth Weise and
Liz Szabo, "Toxic Legacy: Can a Plastic Alter Human Cells?"
USA Today, October 30, 2007, www.usatoday.com.

162 *the final NTP draft report* . . . See the NTP CERHR website:
cerhr.niehs.nih.gov/chemicals/bisphenol/bisphenol.html.

162 *However, what the current science* . . . Within days of the release
of the NTP draft, Senate Democrats introduced a bill to ban
bisphenol A from all products made for infants and children up
to age seven. The legislation, sponsored by Senator Charles
Schumer of New York, also directs the CDC to study bisphenol A
health risks to both children and adults. "There have been
enough warning signs about the dangers of this chemical that we
cannot sit idly by and continue to allow vulnerable children and
infants to be exposed," said Schumer. See Lyndsey Layton, "Senators Propose Ban on Chemical in Plastics," April 30, 2008, *The
Washington Post*, A4. With the future of one of the chemical industry's major products on the line, the American Chemistry
Council worked feverishly to minimize damage. It blamed the
media for "unnecessarily confusing and frightening the public."
But the ACC's steadfast refusal to acknowledge the growing reality of bisphenol A hazards shredded its credibility with reporters
and the public. The ACC held a telephone press conference on
April 17, 2008, defending bisphenol A and calling on the FDA to
update its review of the substance. Only four reporters had questions for ACC representatives, and one journalist wanted to know
if manufacturers were changing anything to make their bisphenol
A–containing products safer. Steve Hentges's answer: "I don't
know how they can be safer than safe. The science does support

the safety." To read the NTP's final September 2008 brief and an informative Q&A on bisphenol A, go to www.niehs.nih.gov/ news/media/questions/sya-bpa.cfm. On December 15, 2008, the Food and Drug Administration said it had no plans to change its assessment of bisphenol A, even after sharp criticism by its own scientific advisers for ignoring available information about health risks. See Lyndsey Layton, "FDA Will Continue to Study Chemical," December 16, 2008, *The Washington Post*, www.washington post.com/wp-dyn/content/article/2008/12/15/AR2008121 5029 20_pf.html.

7. Out of the Frying Pan and onto the Paper: Perfluorinated Chemicals

163 *In 2002, the EPA . . .* The EPA's home page on PFOA and fluorinated telomers contains links to basic information, the agency's voluntary agreements with industry, and a draft risk assessment. www.epa.gov/oppt/pfoa.

163 *The EPA, in a sharply worded . . .* See Juliet Eilperin, "DuPont, EPA Settle Chemical Complaint," *Washington Post*, December 15, 2005, D3.

164 *Outside of the courts . . .* See DuPont's position on PFOA and related issues at www.dupont.com/PFOA/en_US/.

165 *In 1997 . . .* Minnesota Public Radio produced a fascinating series on the legacy of perfluorinated chemicals, news.minnesota .publicradio.org/features/2005/02/22_edgerlym_3mscience.

166 *Seeking answers . . .* The Environmental Working Group reported its findings from the EPA docket in Chemical Industry Archives, www.chemicalindustryarchives.org/dirtysecrets/scotch gard/1.asp.

166 *Indeed, on the day . . .* The e-mail, written by the EPA's Charles Auer, is at www.chemicalindustryarchives.org/dirtysecrets/ scotchgard/4.asp.

168 *If you visit . . .* Teflon World has been removed from www .teflon.com. (Too bad; it was fun.)

168 *During 2004, an estimated 72,000 tons . . .* See "Report of an OECD Workshop on Perfluorocarboxylic Acids (PFCAs) and Precursors," June 18, 2007, Organisation for Economic Cooperation and Development. appli1.oecd.org/olis/2007doc.nsf/ linkto/env-jm-mono(2007)11.

168 *Significantly, data suggests that Mother Nature . . .* Testimony of Scott Mabury, Proceedings of the Standing Senate Committee on Energy, the Environment and Natural Resources, Canadian Senate, February 8, 2007.

169 *Later, according to Richard Wiles . . .* This is "How Teflon Got Stuck: A Policy Analysis Call," Collaborative on Health and the Environment, February 23, 2006, at www.healthandenvironment .org/articles/partnership_calls/363.

170 *After investigating, the EPA agreed . . .* "EPA Settles PFOA Case Against DuPont for Largest Administrative Penalty in Agency History," EPA News Release, December 14, 2005.

170 *For decades, scientists . . .* Telephone interviews with Scott Mabury conducted in 2007.

171 *Lacking statutory authority . . .* See the EPA's PFOA website at www.epa.gov/oppt/pfoa.

172 *Finally, announcing it . . .* See the Federal Register notice at www.epa.gov/EPA-Tox/2006/March/Day-07/12152.nfm.

173 *Yet the news did nothing . . .* See DuPont Shareholders for Fair Value at www.dupontshareholdersalert.org.

174 *The paucity of information . . .* See *DuPont Comments: Response to Petition to Expedite Consideration of PFOA Under Proposition 65*, November 16, 2006, at www.oehha.ca.gov/prop65/public_ meetings/pdf/PFOAPresentationall121206.pdf.

174 *As part of a class-action settlement . . .* Follow the PFOA study at www.c8sciencepanel.org.

174 *In the meantime, DuPont spent $1 million . . .* See See DuPont's product safety information at www2.dupont.com/PFOA/en_ US/product_safety/consumerproduct.html.

174 *And in 2006, the company announced that results . . .* See press release "DuPont Concludes Washington Works Employee PFOA Study," October 17, 2006, at vocuspr.vocus.com/Vocus PR30/Newsroom/Query.aspx?SiteName=DupontNew&Entity =PRAsset&SF_PRAsset_PRAssetID_EQ=103587&XSL=Press Release&Cache=False.

175 *The Society of Toxicologists held . . .* See above, "Perfluoroalkyl Acids: What Is the Evidence Telling Us?" *Environmental Health Perspectives*, Volume 115, Number 5: A344 (May 2007) at www.ehponline.org/members/2007/115-5/focus.html.

175 *As for DuPont's worker health study . . .* Ken Ward, Jr., "DuPont Distorted C8 Study, Scientists Say," *The Charleston Gazette*, October 14, 2007.

176 *DuPont's actions all seem to fit . . .* Paul D. Thacker, "The Weinberg Proposal," *Environmental Science & Technology Online*, February 22, 2006. See commentary at www.vanityfair.com/ online/daily/2008/04/uncovering-the.html.

176 *DuPont is quick to point out . . .* From DuPont website at

www.teflon.com/NASApp/Teflon/TeflonPageServlet?pageId=/
consumer/na/eng/news/news_detail.teflon_history.html.

176 *Plunkett eventually devised a way* . . . Testimony from Joe
Schwarz, director of the McGill University Office for Science and
Society, Canadian Senate, Proceedings of the Standing Senate
Committee in Energy, the Environment and Natural Resources,
February 15, 2007.

177 *Whatever these costs* . . . DuPont E. DeMours & Co., 10-Q Securities and Exchange Commission filing, October 29, 2008.

178 *Because of such concerns, Canada recently* . . . Rebecca Renner,
"Do perfluoropolymers biodegrade into PFOA?" *Environmental
Science & Technology* 42, no. 3 (February 2008): 648–50.

178 *At the heart of this whole mess* . . . Christopher Lau et al., "Perfluoroalkyl acids: A review of monitoring and toxicological findings," *Toxicological Sciences*, published online May 22, 2007.

180 *That perfluorinated pollutants can be measured* . . . My understanding of Mabury's theory of the environmental fate of perfluorinated compounds comes from my personal interviews and
correspondence with Mabury, and from the testimony of Scott
Mabury, Proceedings of the Standing Senate Committee on Energy, the Environment and Natural Resources, Canadian Senate,
February 8, 2007.

182 *Of course, scientists and regulators* . . . Personal interview and
correspondence with Christopher Lau, 2007.

183 *In fact, what FDA researchers found* . . . Rebecca Renner, "It's in
the Microwave Popcorn, Not the Teflon Pan," *Environmental
Science & Technology* 40, no. 1 (January 2006): 4. Study: T. H. Begley et al., "Perfluorochemicals: Potential sources of and migration
from food packaging," *Food Additives & Contaminants* 22, Issue
10 (October 2005): 1023–31.

184 *In 2007, Tim Begley* . . . Rebecca Renner, "PFOA in People:
Food wrappers may be an important, overlooked source of perfluorochemicals in humans," *Environmental Science & Technology*
41, no. 13 (July 2007): 4497–500. Study: Jessica D. D'eon et al.,
"Production of Perfluorinated Carboxylic Acids (PFCAs) from
the Biotransformation of Polyfluoroalklyl Phosphate Surfactants
(PAPS): Exploring Routes of Human Contamination," *Environmental Science & Technology* 41, no. 13 (May 2007): 4799–805.

184 *Consumer groups* . . . Information about the Ohio Citizen Action campaign at www.ohiocitizen.org.

185 *As a practical matter* . . . Sara Schafer Munoz, "EPA Probes
Safety of Key Chemical in Teflon," *The Wall Street Journal*, Janu-

ary 31, 2006. Also: S. Fields, "Another Fast Food Fear," *Environ-mental Health Perspectives* 111, No. 16 (December 2003): A872, at www.ehponline.org/docs/2003/111-16/forum.html.

185 *Unfortunately, there's new evidence suggesting . . .* Liz Szabo, "Non-stick Chemicals May Cut Birth Weight," *USA Today*, August 22, 2007, at www.usatoday.com. These studies are: Benjamin J. Apelberg et al., "Cord Serum Concentrations of Perfluo-roocatane Sulfonate (PFOS) and Perfluorooctonoate (PFOA) in Relation to Weight and Size at Birth," *Environmental Health Perspectives* 115, No. 11 (November 2007): 1670–76, at www.ehp online.org/docs/2007/10334/abstract.html; and Chunyuam Fei et al., "Perfluorinated Chemicals and Fetal Growth: A Study within the Danish National Birth Cohort," *Environmental Health Perspectives* 115, No. 11 (November 2007): 1677–82, at www.ehp online.org/docs/2007/10506/abstract.html.

186 *In test animals, higher doses . . .* "Perfluoroalkyl Acids: What Is the Evidence Telling Us?" *Environmental Health Perspectives* 115, No. 5 (May 2007): A344, at www.ehponline.org/members/ 2007/115-5/focus.html.

186 *One reason is excessive . . .* Testimony of Richard Purdy, free-lance toxicologist, Proceedings of the Standing Senate Committee on Energy, the Environment and Natural Resources, Canadian Senate, February 20, 2007.

188 *Whether current . . .* The first major study of the effect of per-fluorinated chemicals on human fertility was published in January 2009 in the European journal *Human Reproduction* 1, no. 1 (January 2009): 1–6. It suggests that PFCs might be making it harder for couples to have a baby. The study, which looked at 1,240 Danish women, showed that in a mother's blood higher PFOS levels increased the likelihood of infertility from 70 percent to 134 percent and that higher PFOA levels increased the chance of infertility from 60 percent to 154 percent. The study is available online at humrep.oxfordjournals.org. Dr. Philip Landrigan of the Mt. Sinai School of Medicine in New York told ABC News that the dose-response relationship (meaning where the level of exposure goes up, so does the apparent effect) is important. "When you see that kind of parallel trend, especially for the two PFC compounds they looked at, this is powerful evidence," he said. See www.abc news.go.com/print?id=6753022.

188 *PFOS already appears to be . . .* Rebecca Renner, "Snow Shows Perfluorochemical Source," *Environmental Science & Technology Online*, March 28, 2007. Study: Cora J. Young et al., "Perfluori-

nated Acids in Arctic Snow: New Evidence for Atmospheric Formation," *Environmental Science & Technology* 41, No. 10 (2007):
3455–61.

188 *Furthermore, a 2007 study by the CDC* . . .   See "Spotlight on
Polyfluorochemicals," August 2007, Centers for Disease Control
and Prevention, at www.cdc.gov/exposurereport/pdf/factsheet_
pfc.pdf.

189 *And in Minnesota* . . .   See the Minnesota Pollution Control
Agency's perfluorinated chemicals information page at pca
.state.mn.us/cleanup/pfc/index.html, and 3M's PFOS-PFOA
page at solutions.3m.com/wps/portal/3m/en_US/PFOS/PFOA/
?WT.mc_id=keymatch.

8: Reaching Ahead: New Policies

190 *Under REACH* . . .   For more background and details, go to
the European Commission's comprehensive REACH website,
ec.europa.eu/environment/chemicals/reach/reach_intro.htm.

191 *After categorizing* . . .   For more background about the focus on
health and the environment in Canada's chemical management
plan, see www.chemicalsubstances.gc.ca.

193 *Instead, the EPA was championing* . . .   See the EPA's High Production Volume (HPV) page, www.epa.gov/hpv.

194 *"Chemical safety can't be based"* . . .   *Toxic Ignorance: The Continuing Absence of Basic Information About Top-Selling Chemicals
in the United States*, Environmental Defense Fund, 1997, www
.edf.org/documents/243_toxicignorance.pdf.

194 *"As of mid-2007, the industry had not bothered"* . . .   "High Hopes,
Low Marks: A Final Report on the High Production Volume
Chemical Challenge," *Environmental Defense*, July 2007, www
.edf.org/documents/6653_HighHopesLowMarks.pdf.

196 *Over the years* . . .   *Chemical Regulation: Options Exist to Improve EPA's Ability to Assess Health Risks and Manage Its Chemical Review Program*, June 2005, GAO-05-458, www.gao.gov.

197 *When I reached Train* . . .   Telephone interview with Russell
Train, December 2006.

197 *For that kind of background* . . .   Telephone interviews with Terry
Davies, December 2006.

198 *Jim Gulliford* . . .   See Melissa Lee Phillips, "Obstructing Authority: Does the EPA Have the Power to Ensure Commercial
Chemicals Are Safe?" *Environmental Health Perspectives* 114,
issue 12 (December 2006): A706–A709, www.pubmedcentral.nih
.gov/articlerender.fcgi?artid=1764141.

200 *In Senate testimony* . . . Testimony Before the Committee on
Environment and Public Works, U.S. Senate, GAO-06-1032T,
www.gao.gov. Transcript: epw.senate.gov/hearing_statements
.cfm?id=260423.

202 *The man behind* . . . Interviews with Michael Wilson conducted
by telephone and in person in 2006, 2007 and 2008.

203 *Wilson and his coauthors* . . . See Michael P. Wilson et al., *Green
Chemistry in California: A Framework for Leadership in Chemicals
Policy and Innovation*, California Policy Research Center, 2006,
www.ucop.edu/cprc/documents/greenchemistryrpt.pdf.

205 *In their book* . . . Paul T. Anastas and John C. Warner, *Green
Chemistry: Theory and Practice* (New York: Oxford University
Press, 2000).

206 *The Presidential Green Challenge Awards* . . . A list of winners to
1996 is available at www.epa.gov/greenchemistry.

206 *The product, called NatureWorks PLA* . . . See Wal-Mart's website for all of its sustainability programs. The press release for Nature Works is at www.walmartstores.com/FactNews/Newsroom/
5412.aspx.

206 *Businesses intent on scouring* . . . For more details and to see
Tim Greiner et al., *Healthy Business Strategies for Transforming
the Toxic Chemical Economy*, Clean Production Action, June 2006,
go to www.cleanproduction.org/Home.php.

208 *A committee of experts* . . . See *Sustainability in the Chemical Industry: Grand Challenges and Research Needs—A Workshop Report*, National Academy of Sciences, 2005, www.nap.edu/catalog
.php?record_id=11437.

208 *Talking is cheaper* . . . See *Technology Vision 2020: The U.S.
Chemical Industry Realizing the Vision*, Council for Chemical Research, December 1996.

209 *"REACH will not be free"* . . . Yes, European politicians blog.
See Wallström's blog at blogs.ec.europa.eu/wallstrom.

210 *In 2008, Wilson and the Centers for Occupational* . . . See *Green
Chemistry: Cornerstone to a Sustainable California*, The Centers
for Occupational and Environmental Health, University of California, 2008, at coeh:berkeley.edu/greenchemistry/briefing.

210 *The answer to that* . . . See "Governor Schwarzenegger Signs
Groundbreaking Legislation Implementing First-in-the-Nation
Green Chemistry Program," September 29, 2008, Office of the
Governor, at gov.ca.gov/index.php?/press-release/10666/.

 **Acknowledgments**

I am a journalist, not a scientist, heavy on the right brain, lighter on the left. I couldn't have written this book without the help of dozens of infinitely patient left-brainers, those scientists who explained their research and painstakingly answered my questions. You met most of them in these pages, though some who helped me equally you did not. To each and every one of them, collectively too numerous to mention, I am grateful.

The insightful scientists who read my chapters for clarity and accuracy and who shared their expertise deserve more than this very big thank you. If something in this manuscript is not clear, it is through no fault of Arlene Blum, Sarah Jannsen, Margaret Reeves, Ted Schettler, or Glenys Webster; the responsibility is mine alone. My friend and first reader Susan Seager deserves special praise for her help and encouragement.

Understanding the intricacies of toxics laws would have been much more difficult—and a lot less fun—if not for the wisdom of dozens of dedicated workers in government and advocacy organizations on both sides of the Atlantic. Their commitment inspired me when I most needed it.

My agent, Ted Weinstein, believed in this project from the moment we first discussed it. Thanks, Ted, for guiding my leap from daily journalism to book writing and for always checking in.

My admirable friend Angela Cara Pancrazio not only took my author photo, for which I thank her very much, but she also taught me what courage is.

To Denise Oswald, my editor, thank you for your deft editing touch and, above all, your unwavering support. To Jessica Ferri,

Matt Kaye, Lisa Silverman, Laura Bonner, Amanda Schoonmaker, Sarita Varma, Jeff Seroy, Cailey Hall, Sarah Crichton, and the entire team at North Point Press/Farrar, Straus and Giroux, please know how much this first-time author values your help and enthusiasm.

To my partner, Patti Ihnat, whose generosity, kindness, and faith in me made it possible to fulfill this dream and publish my first book, the words "thank you" are not nearly enough to describe my heartfelt appreciation.

And to my loving parents, Bob and Elaine Baker, I wish you could have been here to share the journey.

 *Index*

Page numbers from 233 to 254 refer to notes.

acrylic, 36
Acyloids, 37–38
adhesives, 49, 219
Africa, 84
AGC Chemicals/Asahi Glass,
 173
airborne toxics, 20
Alba, 97
Albert, Eddie, 61
alcohol, 25
aldrin, 43
alkaline cleaners, 165
allergies, 122
allergy tests, 21
All Lacquered Up (nail polish),
 96
aluminum, 224
American Academy of Pediatrics,
 121, 132

American Chemical Society, 205
American Chemistry Council
 (ACC), 16, 26, 52, 129, 199;
 on bisphenol A, 142, 149, 152,
 154, 160, 162; Phthalate Esters
 Panel, 130
American Chemistry Society, 205
American Medical Association,
 157
amphibians, development of, 58,
 59, 69, 71, 72, 73, 78, 85, 218
Anastas, Paul, 205–206
Anderson, Betty, 130
androgen, 85, 155
anogenital distance, 108, 109,
 219
antibodies, 187
antifreeze, 54
Apple, 43, 137

Arctic, 42, 49, 124, 179, 180,
    181–82
Arkema Inc., 173
Arlington, Va., 160
aromatase, 65
aromatase enzyme, 85
artificial rubber, 54
asbestos, 49, 201
asthma, 4, 28, 122, 187, 236
AT&T, 54
atmosphere, 124
atrazine, 57–87, 218–19, 240,
    241; cancer and, 69, 81; in
    drinking water, 57, 67–68, 87,
    218–19; EPA's review of, 57–
    58, 62, 69, 70, 72, 73, 74, 75–
    76, 77, 78–80, 82–83, 84, 87,
    218, 240, 241; in EU, 66; frogs
    and, 63–65, 66, 68, 239–42; in
    United States, 63, 66
atrazinefacts.com, 87
atrazinelovers.com, 81
attention deficit disorder, 4, 156
attention-deficit/hyperactivity
    disorder (ADHD), 143
Auer, Charles, 167, 250
Australia, 179
autism, 4, 236
automobiles, 44, 126, 127, 128,
    164, 209
Avalon Natural Products, 95, 97
Avalon Organics, 97
Avent, 147

Baby Bargains, 161
Baby 411, 161
baby bottles, 9, 27, 142, 143,
    147, 160, 161, 223
bacteria, 86
Bailey, John E., 92–93
Barr, Dana, 62–63
Bayer Material-Science, 144
Beaverton, Oreg., 121

Begley, Tim, 184
behavioral system, 4
Belgium, 178–79
Belliveau, Michael, 127
benzene, 40
Bermuda grass, 66
beverage containers, 141, 142,
    143–44, 147, 150
Bhopal, India, 45
bicycle helmets, 143
biomonitoring, 121, 202, 226; of
    author, 17, 18, 19–20, 235; of
    bisphenol A, 149–50, 152; in
    California, 24; by CDC, 3–4,
    13, 17, 19, 21–23, 236; EPA
    and, 24–25; by EWG, 14–15;
    Patton and, 13, 234; PFOA
    and, 183; phthalates and, 94
Biomonitoring Resource Center,
    234
BioScience, 71
birds, 55
Birnbaum, Linda, 23, 114, 117,
    118–19, 128
birth defects, 13, 221, 225; cos-
    metics and, 93; PFOA and,
    164, 170
bisphenol A (BPA), 9, 27, 141–
    62, 211, 213, 222–23, 247–49;
    ACC on, 141–42, 149, 153,
    154, 160, 162; biomonitoring
    studies of, 149–50, 152; in
    blood, 142, 159; in Canada,
    191; cancer and, 143, 150, 154,
    156, 157; EPA on, 145, 156;
    NPT on, 157–60; reference
    dose for, 146; reproductive sys-
    tem and, 151, 153–54; San
    Francisco's ban of, 155
bisphenol-a.org, 152
bladder cancer, 31
bladders, 103
bleach, 36
blood: bisphenol A in, 142, 159;

chemicals in, 21, 22, 191–92;
  lead in, 5; PBDEs in, 18, 116;
  PFOA in, 165, 170, 171, 178–
  79, 184, 188; PFOS in, 165,
  171, 178–79, 188
blood pressure, 26, 103
body burden, 30, 116, 235
body lotions, 99
Body Shop, 95
Bolinas, Calif., 11–12, 30, 234
BornFree, 161
Borrell, Josep, 192
Boxer, Barbara, 101
brain, 17, 155, 156; cancer of, 4;
  lead and, 5
brake-cleaning solvent, 203
Brandt's cormorant, 179
Braun Medical, 206
Brazil, 178
Breast Cancer Fund, 95, 110, 225
breast milk, 13, 46, 179; PBDEs
  in, 14, 119–23, 140
breasts, 155; cancer of, 4, 27, 31,
  81, 122, 143, 150, 154–55,
  156, 159, 222, 228
British Columbia, University
  of, 235
brominated flame retardants, 20,
  123, 124, 136, 138, 156, 195,
  227; see also polybrominated
  diphenyl ether (PBDE)
bromines, 41, 112, 113, 128
Bromine Science and Environmental Forum (BSEF), 113,
  125, 130, 131, 134, 137
Browner, Carol, 165, 249
Brune, Mary, 123
Buchanan, Sandy, 184
Bucher, John, 157, 161
bug control, 213
building materials, 219
bulletproof vests, 16
bullet-resistant barriers, 143
Burger King, 185

Burson-Marsteller, 113, 131
Bush administration (George W.),
  9, 73–74, 76–77, 129, 200;
  REACH and, 52
butter, 184
butylenes, 40
Byrd's Beauty Shoppe, 88

C8, see perfluorooctanoic acid
  (PFOA)
cadmium, 6, 17, 21
California, 94, 98–100, 174, 202,
  203–204, 209; Assembly Bill
  706 in, 136; biomonitoring in,
  24; PBDEs in, 14, 116, 120,
  125; phthalates banned in, 96,
  105, 235; Proposition 65 in,
  96, 164, 172, 177; toxics laws
  in, 211
California, University of, 44, 109,
  210
California, University of, at
  Berkeley, 58, 74, 80, 202, 218
California Department of Toxic
  Substances Control, 210
California Environmental Protec-
  tion Agency, 123
California Pacific Medical Center
  Research Institute, 154
California Pizza Kitchen, 172
California Senate Environmental
  Quality Control, 199
Camelbak, 223
Campaign for Safe Cosmetics,
  95, 96, 97, 101, 102, 103,
  225
Canada, 42, 120, 161, 178, 180–
  82, 186, 190–91, 251, 253
Canadian regulators, 9, 161
cancer, 13, 26, 28, 49, 122, 225;
  atrazine and, 69, 81; bisphenol
  A and, 143, 150, 156, 157;
  bladder, 31; brain, 4, 27;

cancer (*cont.*)
  breast, 4, 27, 31, 81, 122, 143,
    150, 154–55, 156, 159, 222,
    228; cervical, 150; cosmetics
    and, 90, 91, 93, 95; liver, 186;
    lung, 136; ovarian, 27, 122;
    pancreatic, 31, 81, 186;
    prostate, 143, 156, 222; testic-
    ular, 4, 27, 107, 186; tobacco
    and, 136; vaginal, 150; *see also*
    carcinogens
Cancer Research, 155
candy bars, 168
Canisius College, 82
canned foods, 146
Capitol Mall, 237
carbon, 41
carbon atoms, 40
carbon-fluorine bond, 171
carcinogens, 58, 68, 203; in cos-
    metics, 95; PFOA as, 164, 172,
    174, 177; *see also* cancer
Cargill Dow, 206
caribou, 42
carpet padding, 221
carpet pads, 113
carpets, 6, 15, 16, 19, 112, 116,
    164, 168, 192, 207, 218
Carr, Jim, 83
Carson, Rachel, 35, 55–56, 171;
    criticism of, 239
car trim, 113, 125
Case Western Reserve University,
    151
cast iron, 224
Catholic Healthcare West, 206
cats, hyperthyroidism in, 117–
    18
Cavalcade of America, 54
Census Bureau, U.S., 5
Center for Bioenvironmental
    Research, 239
Center for Environmental
    Health, 225–26

Center for International Environ-
    mental Law, 195, 210, 226
Center for Regulatory Effective-
    ness (CRE), 77–78, 241
Center for the Evaluation of
    Risks to Human Reproduction,
    157
Centers for Disease Control and
    Prevention (CDC), 121, 215;
    address of, 229; biomonitoring
    by, 3–4, 13, 17, 19, 21–23,
    236; on DES, 150; National
    Center for Environmental
    Health, 25; on phthalates, 104
Centers for Occupational and
    Environmental Health, 203
central nervous system, 5
Century of Progress Exposition
    (1933), 53
cervix, 150
chalk, 89
Chapel Hill, N.C., 156
Chapin, Robert, 160
Charleston, S.C., 115
Charleston Gazette, 175–76
Chemical Industry Archives, 250
Chemical Industry Council of
    California, 211
chemical processing, 164
Chemicals, Health and Preg-
    nancy study, 235
Chemicals Inspectorate, 119
Chemtura, 124
Chevrolet, 39
Chicago, Ill., 53
Children's Environmental Health
    Network, 226
China, 5, 140, 164, 179
chlordane, 43
chlorinated flame retardants, 136
chlorinated solvents, 43
chlorine, 41
chlorofluorocarbons (CFCs), 17,
    50, 188

chocolate, 184
chromium, 50
Chrysler, 54
Cia Specialty Chemicals, 173
Ciba, 67
Cincinnati, University of, 147
Clariant Corp., 173
cleaning solutions, 54
Clean Production Action, 207, 226
Cleveland, 151
Clinique, 94
Clinton, Bill, 76
clothes, 224
clothing, 15, 45, 168
coal tar, 39–40
Colangelo, Aaron, 83
Cold War, 33–34, 35
Collaborative on Health and the Environment, 226
Collegeville, Pa., 148
Colombia, 178
Commerce Department, U.S., 52, 111
Commonweal, 12, 156, 234
computers, 15, 16, 43–44, 190, 222; PBDEs and, 113, 125
ConAgra Foods, 172, 185
Concannon, C. C., 111
concealer, 110
congeners, 115, 129
Congress, U.S., 56, 198, 215, 217; cosmetics and, 101; CPSC and, 10, 234; EPA and, 8, 46, 47
construction, 39, 164
Consumer Product Safety Commission (CPSC), 6, 9–10
Consumer Product Safety Improvement Act, 10, 104, 234
cookware, 15
Cookware Manufacturers Association, 178
coolants, 115

Cooper, James, 200
copy paper, 168
corn, 66, 68, 146, 206
Corvettes, 39
Cosmetic, Toiletry, and Fragrance Association (CTFA), 91
Cosmetic Ingredient Review, 91, 93, 94–95
cosmetics, 9, 15, 21, 88–111; birth defects and, 93; cancer and, 90, 91, 93, 95; in EU, 90–91, 93, 95, 96, 97, 100, 104, 105, 110, 190, 242; FDA and, 91–92, 94, 95, 101–102, 105; phthalates in, 21, 93–95, 104, 105–106; reproductive problems and, 93, 95; in U.S., 90
cosmeticsdatabase.com, 103
cosmeticsinfo.org, 102
Costco, 140
cotton, 45
couches, 222
Council for Chemical Research, 208
Cover Girl, 94
Crompton, 124
crops, 55
cryptorchidism, 107
cynomolgus monkeys, 166
cysts, 150

Daikin, 173
Dairkee, Shanaz, 155
dairy, 20, 221
Dalton Carpet Outlet, 173
Data Quality Act, 76, 77–78
Davies, Terry, 197–98, 200
DDT, 17–18, 20, 36, 38, 46, 50, 55, 67, 121, 171
Deca, 125, 126–27, 128–29, 130–39, 221
deethylatrazine, 86
Dell, 137, 138

Del Laboratories, 96
Democratic Party, U.S., 7
dental floss, 164
denture cleaners, 165
deodorants, 21, 90
detergents, 104, 220
developmental system, 4, 118,
    175
device regulation, 9
Devine, Jon, 77–78
Devon Ice Cap, 181
diabetes, 4, 27, 122, 143, 156,
    187, 222, 236
dibutyl phthalate (DBP), 94, 95,
    96, 106, 108, 130
dieldrin, 43
diet, 25
diethylhexyl phthalate (DEHP),
    94, 95, 107, 158, 220
diethyl phthalate (DEP), 95
diethylstilbestrol (DES), 150
Dillard's, 173
Dinan, Frank J., 82
dioxins, 17, 20, 50, 121, 128, 156
Disneyland, 33, 34, 35–36, 39,
    55, 237
Ditz, Daryl, 195–96, 210–211
dog beds, 168
Dole, Bob, 82
dolphins, 179
Douglas B-19, 37
Dow, 16, 36
Dow Chemical, 144, 158
drapes, 112, 113, 168
Dr. Brown's, 147
drug, 9
drywall, 39
Duff, Robert, 132, 133, 134, 136
Duke University, 130
DuPont, 16, 19, 36, 37, 53–54,
    163–64, 167, 168, 170, 172,
    173–74, 175–78, 180, 187,
    188, 189; Epidemiological Re-
    view Board, 175

DuPont Shareholders for Fair
    Value, 173
dust, 213, 218, 219, 222; PBDEs
    in, 115, 117
dyes, 40, 49, 54

Eagleton, Thomas F., 101
Ecorisk panel, 240
Eddie Bauer, 172
eggs, 60, 65, 151, 152, 221
Egypt, 89
electrical equipment, 115
electric lights, 55
electrofluorination, 168
electronics, 164, 221
Ellison, Keith, 87
endocrine system, 60–61, 239;
    chemical disruption of, 4, 8,
    26–27, 58, 59, 62, 66, 68, 72,
    77, 97, 117, 142, 146, 222; see
    also hormones
endometriosis, 4, 151
endpoints, 193–94
endrin, 43
Environmental Defence
    (Canada), 161
Environmental Defense, 193–94,
    211, 226
Environmental Health Perspec-
    tives, 77, 85, 108, 116, 185
Environmental Health Sciences,
    29
Environmental Health Strategy
    Center, 127
Environmental Health Studies, 62
Environmental News Service, 73
Environmental Protection Act,
    186
Environmental Protection
    Agency (EPA), 23, 35, 117,
    128, 129, 135, 187, 192, 200;
    ACC and, 52; asbestos banned
    by, 201; atrazine reviewed by,

57–58, 62, 69, 70, 72, 73, 74, 75–76, 77, 78–80, 82–83, 84, 87, 218, 240, 241; biomonitoring and, 24–25; on bisphenol A, 145, 156; budget of, 200; chemicals banned by, 49; chemicals registered with, 37, 43; confidentiality claims and, 47–48; Congress and, 8, 46, 47; Experimental Toxicology Division, 114; on fluorinated telomers, 249; Health and Environmental Effects Research Laboratory, 183; inadequacies of, 8, 16, 48–49, 50–51; National Pollution Prevention and Toxics Advisory Committee (NPPTAC), 50; Office of Pollution Prevention and Toxics (OPPT), 50, 167, 192–93, 196–99, 200; Office of Prevention, Pesticides and Toxic Substances, 7, 46, 185; Office of Water, 219; PBDEs studied by, 115, 118, 129, 130; PCBs banned by, 17–18, 20, 49, 124; on PFBS, 188; on PFOA, 14, 19, 163–64, 165, 167, 170, 171–72, 173, 174–75, 176, 178, 183, 185, 249; on PFOS, 19, 165, 166–67, 171–72; phthalates and, 109, 243; Presidential Green Challenge Awards of, 206; Teflon reviewed by, 163–64
Environmental Science & Technology, 176, 184
Environmental Working Group (EWG), 14–15, 28, 93, 103, 130, 147, 158, 166, 169–70, 227, 235, 243, 250
Environment California Research & Policy Center, 147, 226
Environment Canada, 60, 115

Environment Health News, 226–27
Environment Health Strategy Center, 227
epididymis, 150
epoxy resins, 141, 144, 222
Estée Lauder, 94
ester bonds, 147
estrogen, 65, 85, 142, 150, 153, 155
ethylene, 40
ethylene vinyl acetate (EVA), 220
Ethyl Gasoline Corp., 53
European Chemicals Agency, 194
European Commission, 51, 135, 138, 210, 253
European Court of Justice, 138
European Food Safety Authority, 149
European Parliament, 138, 190, 192, 209
European People's Party, 192
European Union, 190–91; chemical industry in, 52; chemical regulations in, 5, 14, 29, 51, 66, 125, 134–35, 138, 190, 191–92, 199, 200–202, 214; cosmetics in, 90–91, 93, 95, 96, 97, 100, 104, 105, 110, 190, 242; DBP in, 96; *see also* REACH
Evenflo, 147
exercise, 26
Exponent, 130
eyeglass lenses, 143
eyeliner, 98
eye makeup, 90
eye shadow, 110

face makeup, 90
face paints, 99
fats, 89, 176

fatty acid metabolism, 175
fax machines, 125, 126
feces, 13
Federal Insecticide, Fungicide,
    and Rodenticide Act, 58
Federal Register, 251
Federal Reserve, U.S., 44
Feinstein, Dianne, 101, 104, 105
fetuses, 153, 159, 160, 185, 221
firefighting foams, 165
Firemaster, 550
fish, 20, 55, 70, 85, 189, 221
Fish and Wildlife Service, 82
flame retardants, 6, 16, 213;
    brominated, 20, 123, 124, 136,
    138, 156, 195, 227; chlori-
    nated, 136; see also polybromi-
    nated diphenyl ether (PBDE)
floor coverings, 220
floor polishers, 165
flowers, 55
fluorinated telomers, EPA on, 249
fluorine, 176, 187
fluorocarbons, 185
fluoropolymers, 18, 164, 168,
    170, 173, 187–88
fluorotelomers, 167–69, 170,
    178, 185, 188, 213
food, 9, 15
Food, Drug, and Cosmetic Act
    (1938), 89, 90, 100, 110, 242
food containers, 6, 15, 16, 21,
    184, 213; bisphenol A in, 141,
    142, 143–44, 147, 150
Food and Drug Administration
    (FDA), 5, 9, 72, 157, 250; cos-
    metics and, 91–92, 94, 95,
    101–102, 105; PFOA and,
    174–75, 183
food packaging, 18, 146–47, 168,
    184–85, 190, 219, 224, 252
Ford, 54
Ford, Sherrie, 66
formaldehyde, 96

Formica, 36
Fort Worth Star-Telegram, 137
fossil fuels, 208
Foster, Paul, 106
foundation, 99, 110
FOX News, 72
fragrances, 104, 220
Freedom of Information Act, 73,
    74, 92
frogs, 58, 60, 63–65, 66, 68, 70,
    84, 86, 239–42; hermaphro-
    ditism in, 63, 64, 66, 84;
    larynxes of, 70, 73; see also
    amphibians
frying pans, 37, 178
Fuller, Irving "Pep," 48
fungicides, 64
furans, 20, 128
furniture, 6, 15, 19, 45, 112, 168,
    207, 221, 224; PBDEs and,
    113, 117, 125, 126

Gap, 172
garden hoses, 219
Garske, Lynn, 207
gasoline, 206; lead in, 5, 24–25,
    53, 112
Geigy, J. R., 67
Geigy Chemical Corporation, 67
Geiser, Ken, 202
General Accounting Office,
    U.S., 8
General Electric, 53, 54, 144
General Motors, 54
genes, 25, 26, 28, 29, 32, 155–56,
    236
genetically modified foods, 35
Geological Survey, U.S. (USGS),
    57, 67, 68, 86–87, 218
Gerard, Leo, 173
Gerber, 147
Germany, 178, 179
GE Symphony Hour, 53

Gilliom, Robert, 86, 87
global warming, 4
Globe and Mail (Toronto), 161
globulin, 153
glues, 45
Goldman, Lynn, 7–8, 46–47, 48, 200
golf courses, 66
"Good Genes Gone Bad" (Myers), 155–56
Gore, Al, 4–5
GORE-TEX, 164, 169, 173, 182, 223
Government Accountability Office (GAO), 8, 16, 47, 48, 196, 235
Graduate, The (film), 41
Grandjean, Philippe, 28
Granholm, Jennifer, 208
grass, 66
grease, 223
Great Depression, 53, 111, 209
Great Lakes Chemical, 124
Greece, 89
Green Acres (TV show), 61
green beans, 146
Green Chemistry: Theory and Practice, 205
green chemistry, 205–206
Green Chemistry Institute, 205
Greenpeace, 210, 227
Gregoire, Christine, 105
Grist, 123
guava, 66
Gulf of Mexico, 180
Gulliford, Jim, 198
Guth, Joe, 50

Hain Celestial Group, 98
hair, 13
hair color, 90
hair relaxers, 110
halogens, 41

Hammond, Michele, 116
Hammond Holland, Rowan, 116
Hansen, Sally, 96
Harper's, 38
Harris poll, 5
Harshaw Chemical, 37
Harvard School of Public Health, 28
Harvard University, 58, 61
Hauschka, Dr., 99
Hayes, Kassina, 61
Hayes, Tyler, 61
Hayes, Tyrone, 57, 58–59, 60–61, 63, 69–71, 72, 77, 78–82, 84, 86, 218, 239, 240; Steeger's letter to, 74–76
Hazen, Susan, 185
Health Care Without Harm, 227
Health Services Department, 100
heart disease, 26
Heindel, Jerry, 157
Henry, Carol, 26
Hentges, Steve, 149, 154, 160
heptachlor, 43
herbicides, 63, 64, 68
Herman Miller furniture, 39, 207
hermaphroditism, 63, 64, 66, 84
Hewlett-Packard, 137, 138
hexachlorobenzene, 43
hexavelent chromium, 50
Hexion Specialty Chemicals, 144
Hickox, Winston, 136
High Production Volume (HPV) Challenge, 193–94, 198
Hill, Austin Bradford, 136
Hoffman, Sasha, 99
Holland, Jeremiah, 116
Holliday, Charles O., Jr., 173
Hooker Electrochemical, 37
Hooper, Kim, 123, 125
Hoppin, Jane, 106
hormone replacement, 103

hormones, 239; chemical disruption of, 4, 13, 26, 91, 108; growth, 61; of vertebrates, 59–61; *see also* endocrine system
Houlihan, Jane, 14, 28, 130
House Committee on Energy and Commerce, 149
House Committee on Oversight and Government Reform, 8
household cleaners, 15
Howard, Frank, 53
Hoyle, William, 149
Hudson Institute, 239
hula hoops, 36
human fetal programming syndrome, 175
Human Toxome Project, 235
Hunt, Pat, 151–53
hydraulic fluids, 165
hydrocarbons, 40
hydrochloric acids, 41
hydrogen, 176
hyperthyroidism, 117–18
hypospadias, 107, 109–110
hysterectomies, 151

I Didn't Know Carcinogens Came in Coral, 96
IKEA, 119, 137
Illinois, 87
immune system, 122, 175, 186, 187
India, 179
infant formula, 147
infections, 122
inks, 49
inorganic chemicals, 41
insecticides, 54, 64, 165
insulating fluids, 115
insulations, PBDEs and, 113
insulin, 108, 156, 219
InterfaceFABRIC, 207
Inter-Industry Group for Light Metal Packaging, 149

International Electrotechnical Commission (IEC), 126
International Epidemiology Institute, 185
International Journal of Corporate Sustainability, 140
Inuits, 42
Investor Environmental Health Network, 139
Iowa City, Iowa, 84
IQ, 118
Italy, 178

Jackson, Richard, 25
James, Darren, 110
James, Olivia, 109–110
Japan, 179
J.Crew, 172
Jennings, Bruce, 199
Jirtle, Randy, 236
Johns Hopkins Bloomberg School of Public Health, 7, 48
Johns Hopkins University, 175, 185, 200
Jolie, Angelina, 99
Jones, Enesta, 74
Journal of American Medicine, 89
junkscience.com, 72
Justice Department, U.S., 177

Kaiser Permanente, 207
Kansas Corn Growers Association, 77
Katz, Linda M., 94
Kendall, Ronald, 70
Kennedy, Donald, 72
Kerry, John, 101
KFC, 172
Kim, Katherine, 61
Kiss My Face, 95
Knight Ridder News Service, 72

Korea, 179
Krasnic, Toni, 171, 178
Kyte, John, 125, 126–27, 129,
    137

Lake Michigan, 62
Lancet, 28
Lancôme Paris, 94
Landrigan, Philip, 28
larynx, 70, 73
Lash Lure, 88–89, 242
Lau, Christopher, 183
lawns, 66
Lawrence Berkeley National
    Laboratory, 237
lead, 17, 89, 121; in gasoline, 5,
    24–25, 53, 112; in lipstick,
    102; in toys, 5–6
leather, 50
Lebedev, Greg, 52
Lerner, Michael, 12
Levi Strauss, 172
Lewis, Sanford, 174
lichen, 42
linens, 168
lipids, 103
liposuction, 114
lipstick, 90, 95, 98–99, 102
Liroff, Richard, 139–40
liver, cancer of, 186
livers, 118, 171, 175, 179, 186
livestock, 55
lobbying, 52
Locke, Gary, 244
Long, Tim, 94
L'Oréal, 94
Los Angeles Times, 158
lotions, 110, 220
Love Canal, 45
Lowell Center for Sustainable
    Production, 139, 201, 202
Lowest Observed Adverse Effect
    Level (LOAEL), 145

lubricants, 49, 168
luggage, 168
Lululemon Athletica, 161
lung cancer, 136

Mabury, Scott, 171, 179, 180,
    181, 182, 183, 184, 188, 252
M.A.C., 94, 99–100
macadamia nuts, 66
Maffini, Marcel, 159
Maine, 124, 136
Maine Center for Disease Con-
    trol and Protection, 125–26
Maine Department of Environ-
    mental Protection, 125
Maine Department of Health
    and Human Services, 129
Making our Milk Safe (MOMS),
    123, 227
malaria, 38
Malaysia, 179
Malkan, Stacy, 103
mammals, 70, 85
mammary glands, 155
mammary tumors, atrazine
    and, 57
Manchester, England, 17
Manhattan Project, 37, 177
Marc Lappé, 103
margarine, 184
Marin Civic Center, 98
Marin County Board of Supervi-
    sors, 98
Maryland, 161
mascara, 98, 110
Massachusetts, 208
Massachusetts, University of, 201
Massachusetts General Hospital,
    107–108
maternal phthalate metabolites,
    219
mattresses, 221, 222; PBDEs
    and, 113, 125

Max Factor, 94
Maybelline, 94
McCally, Michael, 21
McDonald's, 172, 185
McEwan, Gerald, 91
McGill University, 251
meal replacements, 146–47
meats, 20, 221
meiosis, 151
melamine dishes, 36
Melon of Troy, 96
menstruation, 61
mercury, 17, 121
metal, 50, 112, 120
metallic compounds, 41
metals, 41
methane, 40
methyl isocyanate, 45
methylmercury, 20–21
mice, 142, 151, 152, 153, 154,
    179
Michigan, 208
microwave popcorn bags, 168,
    183–84, 185, 213, 224
microwaves, 16, 30, 147, 220
milk, 114
Milloy, Steven, 72, 241
mink, 179
Minnesota, 84, 189, 235
Minnesota Pollution Control
    Agency, 253
Minnesota Public Radio, 249
mirex, 43
Miss Chemistry, 54
Mississippi River, 67, 180
Missouri, 60, 69
Missouri, University of, 141
Missouri River, 67
Miss Teen USA World, 99
Miss Treatment USA, 96
Mittelstadt, Katie, 122
Mittelstadt, Laura, 121, 122, 140
Mittelstadt, Nathan, 121, 122
moisturizers, 90, 97

monkeys, 166, 187
Monsanto, 34–36, 39, 55
Monsanto Magazine, 55
Mount Equipment Co-op, 161
Mount Sinai, 13, 28, 234
Müller, Paul, 38
Musser, Hazel Fay, 88–89
Myers, John Peterson "Pete," 29,
    31–32, 72, 155–56, 236–37
Myller, Riitta, 192

nail polish, 90, 94, 96, 110, 168
nails, 13
Nalgene bottles, 27, 213, 223
National Academy of Sciences, 8,
    47, 208, 243
National Association of Science
    Writers, 35
National Center for Case Study
    Teaching in Science, 82
National Children's Study, 231
National Environmental Trust,
    227–28
National Institute of Environ-
    mental Health Sciences
    (NIEHS), 27, 85, 106, 150,
    153, 156, 157; address of, 230;
    see also National Toxicology
    Program (NTP)
National Institutes of Health,
    156, 157; Genes, Environment
    and Health Initiative of, 236
National Institute of Standards
    and Technology, 115
National Press Club, 197
National Research Center for
    Women & Families, 228
National Resources Defense
    Council (NRDC), 73, 82, 83,
    228
National Toxicology Program
    (NTP), 27, 157–62, 188, 250;
    address of, 230

natural gas, 40
Natural Resources Defense
  Council, 18, 77
Nature, 70
NatureWorks PLA, 206
Naugahyde, 36
Nebraska, 85–86
Nelson, Bill, 189
neurological deformities, 13, 28
Newbold, Retha, 150–51
New York Times, 3, 55
New York World's Fair (1939), 54
Ngo, Victoria, 61
n-hexane, 203
Niagara Falls, N.Y., 45
Nixon, Richard M., 197
Nobel Prize, 33, 38
nonmetallic compounds, 41
No Observed Adverse Effect
  Level (NOAEL), 145
Nord, Nancy, 10
Nordstrom, 172
Northwest Coalition for Alterna-
  tives to Pesticides, 228
Northwest Environment Watch,
  228
Not Just a Pretty Face (Malkan),
  103
Novartis, 58, 67; *see also* Syngenta
nuclear war, 33
Nudelman, Janet, 95, 96, 110
number 7 stamp, 223
nylon, 36
nylon seat belts, 16

Oakland Tribune, 74, 116
obesity, 4, 27, 108, 143, 156, 219,
  222, 236
Octa, 125, 127–28, 133, 136, 221
Office of Enforcement and Com-
  pliance Assurance, 73–74
Office of Environmental Health
  Assessments, 132

Office of Environmental Quality,
  198
Office of Pollution Prevention
  and Toxics (OPPT), 50, 167,
  192–93, 196–99, 200
Office of Prevention, Pesticides
  and Toxic Substances, 7, 46,
  185
Office of Technology Assessment, 8, 47
Office of the U.S. Trade Representative, 192
Ohio Citizen Action, 184, 252
Ohio River, 67, 180
oil, 40
oils, 104, 176
oocytes, 152
OPI Products, 96
Oregon, 140
organ damage, 2
organic food, 30, 212
Organisation for Economic Co-
  operation and Development,
  168, 238
organochlorines, 41–42, 43, 238
Orly International, 96
Orville Redenbacher popcorn,
  185
osteoporosis, 122
Our Stolen Future (Myers et al.),
  29, 72, 237
ovarian cancer, 27, 122
ovaries, 65, 151
oxygen, 128
ozone, 17, 50, 188

paints, 15, 50, 54, 168, 220
Panasonic, 139
pancreatic cancer, 31, 81, 186
Papa John's, 172
paper cups, 168
paper plates, 168
paper towels, 147

Pap smears, 21
parabens, 97
Paracelsus, 26
paraphenylenediamine (PPD),
    88
Parkesburg, W.Va., 174
pasta, 146
Pastoor, Tim, 67, 79, 240
Patton, Sharyle, 12–13, 20, 21,
    30, 234
peaches, 146
Pearl River, 179
Pediatrics, 105
Penta, 124, 127–28, 133, 136,
    221, 246
perchloroethylene, 203
perfluorinated chemicals (PFCs),
    18, 156, 163–89, 211, 223–24,
    249–53; in Canada, 191; in
    food wrappers, 168, 184–85,
    190, 252
perfluorinated compounds, 235
perfluorobutane sulfonate
    (PFBS), 188
perfluorocarboxylic acids, 171
perfluorooctane sulfonate
    (PFOS), 18–20, 164–65, 168,
    188–89, 223; biological half-
    life of, 182; in blood, 165, 171,
    178–79, 188; EPA on, 19, 165,
    166–67, 171–72; structure
    of, 171
perfluorooctanoic acid (PFOA),
    168–69, 173–74, 178, 189,
    224; biological half-life of,
    182–83; biomonitoring and,
    183; birth defects and, 164,
    170; in blood, 165, 170, 171,
    178–79, 184, 188; as carcino-
    gen, 164, 172, 174, 177; EPA
    on, 14, 19, 163–64, 165, 167,
    170, 171–72, 173, 174–75,
    176, 178, 183, 185, 249; fatty
    acid metabolism affected by,

175; FDA and, 174–75, 183;
    structure of, 171
perfumes, 90
perineum, 108
Perkin, William Henry, 40
persistent organic pollutants
    (POPs), 13, 20, 42–43
Personal Care Products Council,
    91, 102
Peru, 140
pesticides, 20, 22, 23, 43, 46, 55,
    86, 104, 116, 156, 213, 228; *see
    also* atrazine
Pesticides Action Network North
    America, 228
pet food, 5, 9
petroleum, 40, 97
Pfizer, 160
phonographs, 55
photographic film, 165
photosynthesis, 67
phthalates, 21, 111, 207, 219–20,
    243; California's ban on, 96,
    105, 235; in Canada, 191;
    CDC on, 104; in cosmetics, 21,
    93–95, 104, 105–106; diethyl,
    95; diethylhexyl, 94, 95, 107,
    157–58; EPA and, 109; genes
    and, 156; and male reproduc-
    tive system, 106–110; preg-
    nancy and, 107–110; in toys,
    6, 14, 104–105, 219, 234–35
Phthalates of Pearl, 96
Physicians for Social Responsibil-
    ity, 228
pineapples, 146
Pirkle, Jim, 22, 24–25, 236
pizza boxes, 168
Pizza Hut, 172
placenta, 15, 121
plastic bags, 219
plasticizers, 6, 14, 213; *see also*
    phthalates
plastics, 4, 6, 16, 21, 34–35, 39,

41, 44, 45, 54, 220; global production of, 39; PBDEs and,
112, 114; polycarbonates, 27,
141, 143, 147, 149, 150, 152;
see also bisphenol A
Playtex, 147
Plexiglas, 37, 39
plumbers' tape, 164
Plunkett, Roy J., 176–77, 187–88
Point Reyes National Seashore, 12
Poland, 178
polar bears, 49, 179
polishes, 168
pollution, in bodies, 3–4
polybrominated diphenyl ether
(PBDE), 18, 112–40, 221–22,
244–47; in blood, 18, 116,
178–79; in breast milk, 14,
119–23, 140; Deca, 125, 126–
27, 128–29, 130–39, 235; in
dust, 115, 117; EPA's study of,
115, 118; hyperthyroidism and,
117–18; Octa, 125, 127–28,
133, 136; Penta, 124, 127–28,
133, 136
Polycarbonate/BPA Global
Group, 149
polycarbonate plastic, 9, 143–44,
148–49, 152
polychlorinated biphenyls
(PCBs), 17–18, 20, 43, 46,
114–15, 121, 123–24, 127,
156, 221; U.S. ban on, 17–18,
20, 49, 124
polyester, 36
polyethylene, 36, 206
polyfluoroalkyl phosphate surfactant (PAPS), 184
polymers, 39, 150, 172
polyps, 151
polystyrene cups, 36
polytetrafluoroethylene (PTFE),
see Teflon
polyurethane, 112

polyvinyl chloride (PVC), 6, 43,
206–207, 219, 220, 227
polyvinylidene chloride, 206
popcorn, 168, 183–84, 185, 213,
224
Portland, Oreg., 17, 92
pots, 178
powder, 110
Powers, Lisa, 102
Precursor Alcohol Atmospheric
Reaction and Transport
(PAART), 181, 182
pregnancy, 29, 103, 150, 217;
phthalates and, 107–110
Presidential Green Challenge
Awards, 206
printers, 126
printing inks, 168
*Proceedings of the National Acad-
emy of Sciences*, 70
Procter & Gamble, 94
propylene, 40
prostate, 153, 155
prostate cancer, 143, 156, 222
proteins, 103
PTA, 88
puberty, 4, 26, 143, 221, 222
Purdy, Richard, 164, 165–66,
186–87, 189, 252
Puritans, 89

Radisson Hotel, 199
rats, 142, 154, 159, 166, 186;
atrazine and, 57; phthalates
and, 106–107, 108
REACH (Registration, Evaluation, Authorisation and
Restriction of Chemical
Substances), 51–52, 190,
193–95, 200–202, 209–211;
website of, 253
recycling, 223
refrigerants, 54, 176, 188

refrigerators, 44, 55
reproductive system, 4, 151, 175; cosmetics and, 93, 95
Reproductive Toxicology, 143, 157
reptiles, 70, 85
Republican Party, U.S., 7
Resources for the Future, 198
Revlon, 94
Rice, Deborah, 129
Roberts, Rebecca, 148
robots, 54–55
Rochester, University of, 108
rodents, 85
Rohm and Haas, 38
RoHS Directive, 138
Rome, 89
Roosevelt, Franklin Delano, 89
Royal Society of Medicine, 136
RTI International, 144
Ruff, Victoria, 99
Rug Doctor, 172
Russell, Steve, 199

Safe Cosmetics Act (2007), 100
Safe Drinking Water Act, 57
St. Augustine grass, 66
St. Paul, Minn., 84
salt, 41
Sam's Club, 206
Samsung, 139
Sandoz, 67
San Francisco, Calif., 154, 207
San Francisco Chronicle, 234
San Pablo Bay, 98
Saran Wrap, 36, 206
Sass, Jennifer, 77–78
Schecter, Arnold, 124
Schneider, Ed, 173
Schwarz, Joe, 251
Schwarzenegger, Arnold, 24, 100, 174, 204, 210, 235
Science, 72

Science and Environmental Health Network, 50, 228
Sciences International, 158–59
Scientific Analysis Laboratories, 17
SC Johnson, 206
Scotchgard, 19, 164, 165, 169, 188, 223–24
Sears, 172
secondhand smoke, 72
Securities and Exchange Commission, 177
semen, 13, 60, 143, 179, 222
semiconductors, 164
Senate, U.S.: mandatory toy safety testing wanted by, 5–6; phthalates in, 104
Senate Bill 1313 (California), 174
Senate Committee on Environment and Public Works, 198, 200
September 11, 2001 terrorist attacks, 31–32
sexual development, 70
shampoos, 90, 97, 165
Sharp, 139
shaving gels, 97
shoes, 224
shower curtains, 220
Shriftman, Morris, 97–98
Sigg bottles, 213
Sightline Institute, 228
Silent Spring (Carson), 35, 55–56, 239
Silent Spring Institute, 228
simazine, 67
sippy cups, 142, 149, 223
skin care, 98, 103
Skin Deep, 103, 243
Smith, Rick, 161
Smithsonian Institution, 237
smoking, 25
soaps, 4, 21, 220

Society of Environmental Toxicology and Chemistry, 84
Society of Toxicologists, 175
soda, 146
soil, 124, 189
Solomon, Gina, 18
Solvay Solexis, 173
solvents, 36, 45, 104, 203; see also polychlorinated biphenyls (PCBs)
Sommer, Charlie, 55
Sony, 137, 139
sorghum, 66
soups, 146
South China, 179
sperm, 14, 60, 65, 107, 108, 150, 219, 221
sperm count, 4, 155
stainless steel, 224
Stainmaster, 223
stain repellent, 6, 168, 213, 223, 235
Stalingrad, battle of, 38
Standard Oil, 54
Stanford Genome Technology Center, 154
Stanley Steemer, 172
Stapleton, Heather, 130–31
starch, 89
State Department, U.S., 52
steak, 114
Steeger, Tom, 73, 74–76, 78–79, 241
Steel, U.S., 54
Stinson Beach, 11
Stockholm, 13, 43
Stockholm Convention, 236
stress hormones, 86
sudden infant death syndrome, 122
sulfuric acids, 41
sunburns, 89
Sunoco Chemicals, 144
Survey of Consumer Finances, 44

Swan, Shanna, 108, 109, 130
sweat, 13
Sweden, 119–20, 179
Syngenta, 58, 61, 63, 66–74, 76, 77, 78, 79–80, 82, 83, 84, 87, 240, 241
Synthetic Organic Chemical Manufacturers Association, 200

Taco Bell, 172
Target, 105
Teflon, 19, 37, 156, 163–64, 167, 169, 174, 176–78, 213, 223, 224; EPA review of, 163–64
Teflon World, 168
telephones, 44, 125, 126
televisions, 6, 44, 190, 222; PBDEs and, 14, 113, 126, 139
telomers, 167–69
Tennessee River, 179
testes, 150, 155
testicular cancer, 4, 107, 186
testicular dysgenesis syndrome, 107
testosterone, 65
tetraethyl, 36, 53
Texas, 120, 198, 202
Texas, University of, 124
Texas Centennial Exposition (1936), 53
Texas Tech University, 70, 83
textiles, 164
3M, 164, 165, 166, 167–68, 172, 185, 186, 188–89, 235
3M/Dyneon, 173
3M Scotchgard, 19, 164, 165, 169, 188, 223–24
thymus glands, 187
thyroid, 155, 186, 221, 235
thyroid system, 14, 117–19
Tickner, Joel, 201, 202
Time, 33, 236

Times Beach, Mo., 45
tin oxide, 89
TNT, 40
tobacco, 72, 136, 154
Tokyo, 80
toluene, 40, 96
toothpaste, 90, 97
Toronto, University of, 115, 179
toxaphene, 43
Toxic Ignorance, 194
Toxic Substances Control Act (TSCA; 1976), 7, 16, 46, 47, 49, 123, 135, 170, 193, 195, 196–99, 200, 214, 217, 235
toys, 4; bisphenol A in, 155; lead in, 5–6; phthalates in, 14, 104–105, 219, 220, 234–35
Toys"R"Us, 105
Tozzi, Jim, 77
Trade Representative, U.S., 8
Train, Russell E., 197
trains, 126
tranquilizers, 34
Triazine Network, 77
triazines, 58
Tufts University, 159
tuna, 146
Tupperware, 38, 147
Turner, Dan, 167
Tyl, Rochelle, 144
typhus, 38

Uganda, 84
umbilical cord blood, 170, 179
Union Carbide, 36, 37
United Nations: Environment Programme of, 42, 236; Persistent Organic Pollutants treaty of, 13, 20, 43
United States: atrazine in, 63, 66; chemical industry in, 16, 44, 52, 208–209; consumer in-

debtedness in, 44; cosmetics in, 90; DDT and PCBs banned in, 17–18, 20, 49; failure to protect from chemical hazards in, 4, 190–91, 201; industrial substances registered in, 15; leaded gas banned in, 24; lead-safety standards in, 5; Stockholm Convention not adopted in, 236
United Steelworkers, 172, 174
Un-Petroleum, 97
upholstery, 168
urine, chemicals in, 13, 21, 22
urogenital abnormalities, 143, 222
Ursinus College, 149
uterine fibroids, 151
UV, 26

vagina, 150
Variety, 96
vinyl, 38, 219, 220
viruses, 122
VOC paint, 213
voluntary programs, 6, 7, 14
Volvo, 119
vom Saal, Fred, 141–42, 144, 146, 149, 153, 157, 160, 161
von Eschenbach, Andrew C., 9

Waldman, Bruce, 61
Wall Street Journal, 136
Wallström, Margot, 51, 210
Wal-Mart, 105, 172, 206
Walt Disney, 54
W. Alton Jones Foundation, 241
Ward, Ken, Jr., 175
Warner, John, 205
washing machines, 44
Washington, D.C., 115, 166, 169
Washington, University of, 105

Washington Department of Ecology, 134
Washington Post, 67, 78
Washington State, 6, 105, 132–36, 235
Washington State Nurses Association, 132
Washington State University, 151
Washington Toxics Coalition, 104–105
water, 124, 189, 223; atrazine in, 57, 67–68, 87, 218–19
waterborne toxics, 20
water bottles, 152, 161
water cooler jugs, 143
water-repellent jacket, 14
water repellents, 235
waxes, 168
Webster, Glenys, 235
Weinberg Group, 176
Weiss, Robert, 160
Westinghouse, 54
West Virginia, 163, 170, 177, 179, 180
White Paper on Potential Developmental Effects of Atrazine on Amphibians, 241
Wiles, Richard, 166, 169

Wilson, Michael, 202–205, 210
window cleaners, 168
winter wheat, 66
W. L. Gore & Associates, 172, 173
wolves, 42
wood, 50, 112
wool, 112
World Health Organization, 121
World War II, 37
World Wildlife Fund, 139, 197
Wright, Frank Lloyd, 98
WWF Toxics, 228
Wyden, Ron, 100–101

xylenes, 40

Yale University, 175, 205
Yangtze River, 179
yolk, 65
Young, Cora, 182

Zia Natural Skincare, 95
Zonyl, 168

Printed in the USA
CPSIA information can be obtained
at www.ICGtesting.com
LVHW091131150724
785511LV00001B/69

9 780865 477469